Eroding Soils

The Conservation Foundation is a nonprofit research and communications organization dedicated to encouraging human conduct to sustain and enrich life on earth. Since its founding in 1948, it has attempted to provide intellectual leadership in the cause of wise management of the earth's resources.

Eroding Soils

The Off-Farm Impacts

Edwin H. Clark II
Jennifer A. Haverkamp
William Chapman

Eroding Soils: The Off-Farm Impacts

Cover design by Sally A. Janin
Typography by Rings-Leighton, Ltd., Washington, D.C.
Printed by R.R. Donnelley & Sons Company, Harrisonburg, Virginia

The Conservation Foundation
1717 Massachusetts Avenue, N.W.
Washington, D.C. 20036

Library of Congress Cataloging in Publication Data

Clark, Edwin H., 1938—
Eroding soils.

Includes bibliographies and index.
1. Soil erosion—Environmental aspects.
2. Water—Pollution—Environmental aspects.
I. Haverkamp, Jennifer A. II. Chapman, William, 1955- . III. Title.
QH545.S64C55 1985 574.5'222 85-9619
ISBN 0-89164-086-X

Contents

Foreword

Since water-quality programs were first established, the United States has made some significant strides, especially in cleaning up such conventional pollutants as fecal coliform, dissolved solids, and biochemical oxygen demand. The experience has taught us a great deal. As we learn more about how pollutants behave in the environment and as our detection capabilities improve, we are beginning to appreciate today that we need reinvigorated efforts and new initiatives to tackle the remaining water-quality problems. The goal is to assure adequate water supplies, when and where they are needed, for all the many purposes water serves.

Further progress in water cleanup requires that we confront some difficult problems, among them nonpoint source-pollution—that is, runoff from agricultural lands, construction sites, mining operations, city streets, and the like. Runoff and the pollutants it carries take their toll in waterways polluted by sediment, fertilizers, and pesticides; in reservoirs and lakes filled up through accelerated siltation; in degraded breeding grounds for fish and other aquatic life; in increased costs for dredging harbors; and in many other ways. As more point sources of water pollution come under control, nonpoint sources and their effects are becoming increasingly important concerns. The problem has thus far eluded the nation's water-pollution-control strategy, even though runoff accounts for one-third to one-half of the pollution load in many of the largest rivers, lakes, and estuaries.

Eroding Soils: The Off-Farm Impacts examines the problems caused by soil erosion off the farm and, in so doing, addresses the serious problem of nonpoint-source pollution. This book compiles what is known about the effects of erosion on water quality and provides the most thorough tally ever attempted of how much the problems cost each year—in the neighborhood of $6 billion (1980 dollars), the authors estimate.

Debate during the past session of Congress over reauthorization of the Clean Water Act offers some hope that the problem of nonpoint-source pollution will finally get the attention it warrants in water cleanup. But will the nation's farmers adopt well-understood practices to combat erosion and reduce runoff? Farmers are caught in one of the tightest economic situations of this century, with farm debt and public subsidies to agriculture running at record-high levels. How can we ask them, given their situation, to take on added burdens? Past soil conservation efforts were designed to create economic incentives for farmers to preserve their land and thus responded both to the farmer's distress and to conservation needs. It may be time to apply that approach once more, looking, for example, at opportunities to retire substantial erosion-prone lands permanently from production—perhaps by planting trees, as was done throughout the South in the 1950s; those trees are now reaching maturity.

Eroding Soils also synthesizes a number of long-standing Conservation Foundation interests in agricultural resources, water, and toxic substances control. The Foundation's book *The Future of American Agriculture as a Strategic Resource*, published in 1980, examined the potential vulnerability of agricultural resources, the land, water, soils, and other factors that account for the high levels of agricultural output on which the nation has come to rely. Among the most serious issues was soil erosion. A subsequent book, *Soil Erosion: Crisis in America's Croplands?*, looked at soil erosion in more detail. Then followed two Foundation films, seen by millions, that distilled for a general audience the causes and consequences of erosion.

The Conservation Foundation's Water Program, directed by Dr. Edwin H. Clark II, is researching the water-resource conflicts likely to dominate water policy during the next decade and beyond. Controlling runoff is high on the list of priorities. So, too, is controlling toxic contamination of water supplies, which can result from pesticides and other agricultural chemicals washing off farmland. The control of cross-media pollution, in which toxics move about in air, water, and land, is another area in which the Foundation has been heavily involved. These issues have been covered in depth in the Foundation's report, *State of the Environment: An Assessment at Mid-Decade*.

We are grateful to the Atlantic Richfield Foundation, Exxon Company, U.S.A., The Ford Foundation, and the Rockefeller Brothers Fund, whose support made *Eroding Soils* possible. We also want to acknowledge helpful critiques by several individuals who commented on early drafts: Hope Babcock, Norman A. Berg, Pierre Crosson, Robert Davis, Klaus Flach, James Giltmeier, Maureen Hinkel, Edward LaRoe, Richard Magleby, Carl Myers, Clayton Ogg, Karen Prestegaard, Walter Rittoll, Neil Sampson, Wesley Seitz, Richard Smith, Charles

Terrell, Robert Thornson, and R. D. Wauchopp, as well as many others who provided valuable data and insights. These individuals, however, bear no direct responsibility for the contents of this report.

Finally, we are very grateful to Mary H. Ruckelshaus and Julia L. Doermann, who spent long hours searching for and checking obscure data and references, to Marsha G. White, for assisting in the research and carefully typing many drafts of the report, and to Bradley B. Rymph, for his careful editing of the manuscript.

William K. Reilly
President
The Conservation Foundation
May 1985

Executive Summary

As a result of several recent surveys and analyses, two closely related problems—soil erosion and nonpoint-source water pollution—are receiving increased attention. Surveys by the U.S. Department of Agriculture show that America's lands are eroding at substantial rates—over six billion tons a year—despite the billions of dollars that have been spent for soil conservation. As industries and municipalities succeed in cleaning up their waste-water discharges, nonpoint sources have become the major source of water pollution in many parts of the United States. According to some estimates, nonpoint sources are responsible for as much as 73 percent of the total biochemical-oxygen-demand loadings, 99 percent of the suspended solids, 83 percent of the dissolved solids, 82 percent of the nitrogen, 84 percent of the phosphorus, and 98 percent of the bacteria loads in U.S. waterways today. Reports by state water-pollution-control officials identify nonpoint sources as the primary reason that many streams still are not satisfying water-quality standards.

Soil erosion is probably the major cause of such nonpoint-source pollution. One study by the U.S. Fish and Wildlife Service in association with the U.S. Environmental Protection Agency identified nonpoint sources in general as the probable cause of water-quality problems adversely affecting fish life in streams across the United States and agricultural sources in particular as the primary cause of problems in 29.5 percent of the stream miles surveyed. In 17.3 percent of the stream miles, agricultural sources of pollution were considered a "major concern."

Most previous analyses of soil erosion have focused on problems such as diminished agricultural productivity that it causes on the farm. This study focuses on the problems caused by the sediment and other contaminants carried off by storm water after they leave the eroding fields. Between the time they leave the field and the time they finally are carried out to sea or come to rest in a reservoir or some other location, these pollutants can cause a wide variety of in-stream and off-stream damages, influenced by a complex series of hydrological, physical, chemical, and biological interactions. This report attempts to identify and document these various impacts and to estimate their economic costs.

Not all of the impacts identified here are strictly the result of soil erosion. Some are caused by dissolved pollutants that storms might carry off the land even if no erosion occurred. However, because erosion usually is a critical element in the process, no effort has been made to distinguish these impacts from those caused by erosion itself.

Any study such as this must grapple with numerous problems. Many types of damage are only weakly documented. If the damages are well documented, the linkage between them and soil erosion probably is not. And, even if the damages and linkages were known, it can still be difficult to assign economic values to the damages. Thus, the estimates in this study are indicative, not definitive. Figure ES.1 summarizes the estimates for the major damage categories. For each category, a range of costs is given, followed by a single value within that range that represents a best guess of the magnitude of the damages. The final column indicates the amount of these damages that can be attributed to cropland erosion.

Figure ES.1
Summary of Damage Costs
(million 1980 dollars)

Type of impact	Range of estimates	Single-value estimate	Cropland's share
In-stream effects			
Biological impacts		no estimate	
Recreational	950-5,600	2,000	830
Water-storage facilities	310-1,600	690	220
Navigation	420- 800	560	180
Other in-stream uses	460-2,500	900	320
Subtotal—In-stream (rounded)	2,100-10,000	4,200	1,600
Off-stream effects			
Flood damages	440-1,300	770	250
Water-conveyance facilities	140- 300	200	100
Water-treatment facilities	50- 500	100	30
Other off-stream uses	400- 920	800	280
Subtotal—Off-stream (rounded)	1,100-3,100	1,900	660
Total—all effects (rounded)	**3,200-13,000**	**6,100**	**2,200**

Source: Original Conservation Foundation research.

IN-STREAM IMPACTS

In-stream damages are those caused by sediment, nutrients, and other erosion-related contaminants in streams and lakes. They include damages to aquatic organisms, water-based recreation, water-storage facilities, and navigation.

Biological Impacts

Aquatic ecosystems can be seriously affected by sediment and other erosion-related contaminants in complex ways. Sediment can destroy spawning areas, food sources, and habitat as well as directly damage fish, crustaceans, and other aquatic wildlife. Algal growth stimulated by nutrients also blocks sunlight, and pesticides and other contaminants carried off agricultural lands can be directly toxic to fish. The National Fisheries Survey identified agricultural runoff as chronically affecting fish communities in 30 percent of the nation's waters, and fish kill reports have identified such runoff as a major cause of acute episodes.

Although some of these biological impacts are reflected in damage estimates to recreational fishing and commercial fishing, little is known about the overall magnitude of these impacts, and there is no commonly accepted methodology for placing economic values on them. The absence of an estimate of damage costs to biological communities, however, should not be taken as an indication that the impacts are small. They are not. If there were appropriate procedures for placing economic values on these damages, this category might well outweigh any of the others.

Recreational Impacts

Damages to water-based recreational activities is the largest estimated category of in-stream damages. This is not surprising given the economic importance of recreational activities. In 1979, 6.27 percent of disposable income was spent on recreation nationwide, and leisure-time activities accounted for 1 out of every 15 jobs in the country.

All types of water-based recreational activities are likely to be adversely affected by erosion-related pollutants. The value of freshwater fishing is reduced because such related pollutants can reduce fish populations and cause game fish such as trout to be replaced by low-value species. Fishing is also less successful in turbid water because the fish have difficulty seeing the lures. Some of the same problems affect marine recreational fishing. Many marine species reproduce in estuaries or rivers, and, as the deterioration of Chesapeake Bay has demonstrated, they can be severely affected by water-quality problems in these locations.

Boating and swimming are affected by siltation and weed growth physically interfering with recreational activities and by the diminished pleasure associated with boating on or swimming in polluted lakes and rivers. Hunting is affected because many waterfowl depend for their food supply on aquatic vegetation and other aquatic wildlife affected by pollution. The total economic cost of these recreational damages is estimated to be $2 billion per year, of which $830 million can be attributed to cropland.

Water-Storage Damages

Damages to water-storage facilities from sedimentation may become increasingly important because of the increasing cost and diminishing availability of new water-storage capacity in the United States. Approximately one million acre-feet of storage capacity, costing $300 to $700 an acre-foot, is being built annually solely for the purpose of storing sediment.

In some smaller reservoirs, accumulated sediment is periodically dredged from the reservoirs to allow them to continue to function. But, overall, an estimated 1.4 to 1.5 million acre-feet of reservoir and lake capacity is permanently filled each year with sediment. This capacity eventually will have to be replaced, and the construction costs probably will be much higher than the cost of current capacity.

Sediment and nutrients also can affect the rate of evaporation and transpiration from water bodies. Evaporation is a particularly serious problem in arid regions because more than an acre-foot of capacity has to be constructed to provide an acre-foot of yield. Here, suspended sediment and algae may provide a benefit since they can reduce evaporation by reflecting much of the solar energy that would otherwise warm the water in lakes and reservoirs. However, sedimentation and nutrients also can increase the rate of evapotranspiration by stimulating the growth of water-consuming vegetation on lake borders.

The final cost related to water storage is lake cleanup. Lakes are the only type of water body thought to have suffered a net deterioration in water quality over the past decade, and all levels of government are spending substantial amounts for weed control and other cleanup activities. The total annual cost of all these impacts on water-storage facilities is estimated to be $690 million, $220 million of which can be attributed to cropland erosion.

Navigational Impacts

Sedimentation also affects navigation in diverse ways. The major cost appears to be maintenance dredging of harbors and waterways. Other costs include accidents and shipping delays. The cost of these impacts is estimated to be $560 million annually, $180 million of which can be attributed to cropland erosion.

Other In-Stream Impacts

Soil erosion also can damage commercial fisheries and reduce what economists refer to as "preservation values." Commercial fisheries suffer from the same problems as recreational fisheries. Preservation values represent the value that people place on clean water even though they may never make direct use of the water body. Some studies have shown them to be higher than recreational and other user values. The total cost of these two types of damage is estimated to be $900 million, $320 million of which can be attributed to cropland erosion.

OFF-STREAM IMPACTS

Off-stream damages are those that occur before the sediment or contaminants reach a waterway, during floods, or after water is taken from the waterway to be used by industries, municipalities, or farms.

Flood Damages

Flood damages in the United States amount to approximately $5 billion annually. Sediment contributes to these damages in three ways: First, by causing the aggradation of streambeds, it increases the frequency and depth of flooding. Second, because suspended sediment is carried with the flood water, the volume of the water/soil mixture is increased. And, third, many flood damages are caused by the sediment, not the water itself. There may also be some long-term damages to agricultural land if floods leave behind infertile silt. The total of all these damages is estimated to amount to $770 million per year, $250 million of which can be attributed to cropland erosion.

Water-Conveyance Impacts

Sediment also can cause problems in water-conveyance facilities. Some of the sediment settles out in drainage ditches before water actually reaches waterways. Each year in Illinois, for example, highway crews remove from drainage ditches sediment equal in amount to 1.4 percent of the total erosion occuring in the state. The other significant cost for conveyance facilities is the annual maintenance cost of removing sediment and controlling weed growth in the 110,000 miles of irrigation canals in the United States, accounting for 15 to 35 percent of the approximate $300 million in annual canal maintenance costs. The total cost of these damages is estimated to amount to $200 million per year, $100 million of which can be attributed to cropland erosion.

Water-Treatment Costs

The costs of treating water for municipal and industrial uses increase because sedimentation basins must be built, chemical coagulants added, filters cleaned more frequently, and, in extreme cases, activated charcoal filters and other devices added to remove pesticides and other dissolved contaminants. These costs are estimated to amount to $100 million per year, $30 million of which can be attributed to cropland erosion.

Other Off-Stream Impacts

Treating water, however, does not eliminate all the costs to off-stream users. For instance, water-treatment facilities do not remove dissolved salts contributed by agricultural sources. These are estimated to cause $80 million in damages annually to municipal and industrial users in the lower Colorado River basin alone. Nutrients and algae in the water also can cause steam electric power plants to operate less efficiently. However, suspended sediment in cooling ponds may, at the same time, benefit such plants by reflecting heat from the sun.

Farmers using irrigation water that contains sediment and other erosion-related contaminants may also experience increased costs. The fine silt may form a crust on the surface of the soil, reducing infiltration and seed germination. Moreover, sediment and dissolved salts can lower crop yields in other ways as well. However, these costs are partially offset by the value of the nutrients in irrigation water. The net cost of all these other off-stream impacts is estimated to be $800 million per year, $280 of which can be attributed to cropland erosion.

In summary, erosion-related pollutants are estimated to have imposed a net damage cost of $3.2 to $13.0 billion a year in 1980, with the single-value estimate being $6.1 billion per year. Cropland's share is $2.2 billion. The amount of damages caused by sediment alone for all sources is approximately $3.2 to $3.7 billion, with $1.1 to $1.3 billion of that caused by sediment resulting from cropland erosion.

All these estimates should be considered only as order-of-magnitude estimates—that is, the actual costs (or benefits) are more likely to be approximately in the range indicated than they are to be one-tenth as much or ten times as much. However, some potentially very significant impacts—both positive and negative—have not been estimated at all. The costs of biological damages are probably the most important of these.

These estimates pertain to the damages caused by sediment and related contaminants and are not estimates of the benefits that would result from sediment control. For several reasons, the benefits would be expected to be lower than the costs given here. Probably the most important is that, if the sediment entering streams were reduced, the stream banks would tend to erode faster, compensating for that loss. Thus, the sediment concentration in the stream would not be reduced by as much as the external sediment loadings, and many of the damages would continue. Eventually, the stream should return to equilibrium conditions with a lower sediment load, but achieving this new equilibrium might take many years.

POLICY ISSUES

Fortunately, there are a number of soil conservation techniques available which would reduce these off-site damages if widely adopted. Some prevent erosion from occurring in the first place; others intercept sediment and other contaminants before they can cause serious damages. And there is always the possibility of mitigating the damages that do occur.

Most of the first two types of techniques appear to be quite effective at reducing the flow of sediment and associated contaminants into streams. Many can also be implemented at relatively reasonable costs (some actually result in lower production costs), although some, such as improved pesticide and fertilizer management, require improved management skills and effort, the cost of which is difficult to express in financial terms. For the most erosive lands, the only truly cost-effective technique may be totally removing the land from production, for growing row crops such as corn and soybeans.

Many of the techniques also have ancillary effects that may provide additional benefits or costs. Some result in the creation of increased wildlife habitat. Most reduce the amount of storm-water runoff as well as the amount of erosion, decreasing downstream flood flows and increasing groundwater recharge. The increased recharge, however, may cause other problems if it results in the contamination of groundwater by pesticides and nitrates.

The mere fact that effective erosion-control techniques do exist does not mean that they will be adopted by landowners unless government develops a comprehensive, aggressive strategy to reduce off-site impacts. Such a strategy needs to include more than just the education, technical assistance, and cost-sharing programs that have been the staple of traditional soil conservation efforts. Most important, it needs to include a mechanism for "targeting" government and private resources, at both the national and local levels, to controlling erosion and runoff on those lands causing the most serious problems.

An effective strategy probably will also need to include: a more effective set of economic incentives to induce landowners to adopt necessary control techniques; regulatory and enforcement programs to induce particularly recalcitrant landowners to take action; and a mechanism for taking the most erosive lands out of row crop production entirely. There is also a need for additional research on the relative effectiveness of the various control techniques, on what would be the most efficient targeting scheme, and on other selected issues. If a strategy incorporating all these elements is adopted, it should be able to substantially reduce the off-farm impacts of soil erosion without significantly increasing government expenditures. In any case, sufficient information and capacity already exists for the United States to begin immediately to address at least the most serious of these impacts.

1. Muddied Waters

The first Europeans settling in the Piedmont area of the American Southeast in the 18th century found the streams there "transparent," "glittering," and "crystal clear."[1] Colonel William Byrd II of Virginia delighted in the clarity of the Dan River, which was "always confined within its lofty banks, and rolling down its waters as clear as crystal" and where the flakes of mica on the bottom "spangled" the gravel as they reflected the sun.[2]

Within a century, however, the crystal of the area's rivers had been broken, and the spangle was gone. Instead, the rivers were

> in many places filling up with detritus . . . sand and mud . . . which is washed in from the hill-sides so that many shoals are being rapidly obliterated, and at many places where within the memory of middle aged men there were shoals or falls of 5 to 10 feet, at present scarcely any shoals can be noticed.[3]

The problem was the "improvident culture" of the surrounding lands for tobacco and cotton:

> nearly all the lands have been cut down and appropriated to tillage: a large maximum of which have been worn out, leaving a desolate picture for the traveler to behold. Decaying tenements, red, old hills, stripped of their native growth and virgin soil, and washed into deep gullies, with here and there patches of Burmuda [sic] grass and stunted pine shrubs, struggling for subsistance on what was once one of the richest soils in America.[4]

This problem, of course, was neither new nor unique to the United States. The geological history of the Earth is a continual process of uplift and erosion. Over millions of years, the Appalachian Mountains have eroded, with the resultant sediment forming the eastern plains where the Europeans first settled in North America. Eroded materials from much of the central United States have formed alluvial deposits more than 400 feet thick along the lower Mississippi River.[5] Erosion from the Himalaya Mountains has formed deposits probably averaging more than a mile in thickness along the lower Indus River in Pakistan.[6]

Throughout the centuries, however, human actions have substantially accelerated this natural process—often with very unfortunate results. The rich Mesopotamian civilization flourished until its vast irrigation system was destroyed by erosion that resulted when the forests on the surrounding hills were cut down.[7] Syria once supplied lumber, wine, and olive oil for the Roman Empire. But the fertile topsoil eroded, destroying the agricultural potential of the area and leaving much of it a desolate wasteland. In classical Greece, Plato vividly described the effects of erosion even as they existed then: "What now remains compared to what then existed is like the skeleton of a sick man, all the fat and soft earth having wasted away, and only the bare framework of the land being left."[8]

MAGNITUDE OF THE PROBLEM

In the United States, the impacts of erosion were experienced early but often disregarded in a young country that seemingly had a limitless supply of good land and water. As the nation grew increasingly urbanized and industrialized, the water pollution problem was seen primarily as a need to control "point" sources of pollution, such as discharges of municipal sewage or industrial effluents.

Soil erosion was viewed primarily in terms of its potential detrimental effects on agricultural productivity and as a matter of land stewardship. Hugh Hammond Bennett, considered the founder of the soil conservation movement, in 1928 viewed soil erosion as "the biggest problem confronting the farmers of the Nation over a tremendous part of its agricultural lands . . . a little is being done here and there to check the loss . . . an infinitesimal part of what should be done."[9] In the intervening decades, the focus in the United States has remained on the impacts of erosion on farms and on farmers.

Erosion Rates

Not until recently was there solid information on how serious soil erosion was in the United States. Then, in 1977, the U.S. Department of Agriculture conducted its first National Resource Inventory (NRI), which indicated that the annual erosion rate (including nonfarm sources) was 6.42 billion tons, equivalent to about 30 tons per person in the nation, or an average of more than 200 tons of erosion per second. The agency conducted a second NRI in 1982, with results very similar to those in 1977.* Even these results, however, applied only

*The 1982 survey paid much closer attention to the problem of wind erosion than did the 1977 effort, and for this reason came up with higher total erosion rates (although the estimates for water-caused erosion were actually slightly lower than they were in 1977).

to nonfederal lands (about two-thirds of the total U.S. land).

Where does this erosion come from? According to the 1977 survey, the main sources of water-caused erosion were agricultural lands; cropland, rangeland, and pastureland together accounted for 69 percent of the total (figure 1.1). Erosion from streams, gullies, roads, and construction sites accounted for another 22 percent.[10] These estimates, however, provide an incomplete picture of the extent of soil erosion in the United States, since they do not include the substantial erosion that is continually taking place, particularly in the West, as normal geological forces wear down mountains.

To understand the magnitude of these numbers, some comparisons are useful. For instance, Iowa is famous for its corn production. But it also experiences substantial erosion: for every ton of corn raised by an Iowa farmer in 1977, five tons of soil were eroded from the state's cropland.* For every pound of food consumed in the United States

Figure 1.1
Water-Caused Erosion on Nonfederal Land, 1977

	Amount (million tons per year)	Percentage of U.S. nonfederal land)*
Cropland	1,926	38.8
Pastureland	346	7.0
Rangeland	1,155	23.3
Forestland	435	8.8
Streams	553	11.1
Gullies	298	6.0
Roads	169	3.4
Construction sites	80	1.6

* Excluding natural geological erosion (for which no estimates are available)

Source: U.S. Department of Agriculture, Soil Conservation Service, *1980 Appraisal Part I: Soil, Water, and Related Resources in the United States: Status, Conditions, and Trends* (Washington, D.C.: U.S. Government Printing Office, 1981), p. 98; and U.S. Department of Agriculture, *A National Program for Soil and Water Conservation: 1982 Final Program Report and Environmental Impact Statement* (Washington, D.C.: U.S. Department of Agriculture, 1982), p. 9.

*This does not imply that all this erosion results from raising corn. Much of it occurs on lands devoted to other crops.

that year, water caused the erosion of over 22 pounds of soil from all agricultural land.[11] The average rate of water and wind erosion of cropland alone in 1977 was more than 89 tons per second.

Figure 1.2
Annual Sheet and Rill Erosion on Cropland, by Erosion Interval, 1977

Erosion interval (tons per acre)	Total acres (million tons)	Cumulative percentage of acreage	Total sheet and rill erosion (million tons)	Cumulative percentage of erosion
0– 1	131.6	31.8	49.2	2.6
1– 2	74.6	49.8	110.6	8.3
2– 3	51.5	62.3	127.5	14.9
3– 4	35.9	71.0	125.0	21.4
4– 5	26.0	77.3	116.3	27.4
5– 6	17.6	81.6	96.2	32.4
6– 7	12.6	84.6	81.8	36.6
7– 8	9.3	86.9	69.4	40.2
8– 9	7.3	88.7	62.0	43.4
9–10	5.8	90.1	54.6	46.2
10–11	4.8	91.3	50.2	48.8
11–12	3.7	92.2	43.1	51.0
12–13	3.0	92.9	36.9	52.9
13–14	2.8	93.6	37.1	54.8
14–15	2.4	94.2	34.6	56.6
15–20	7.8	96.1	134.8	63.6
20–25	4.4	97.1	98.0	68.7
25–30	2.9	97.8	80.6	72.9
30–50	5.5	99.1	209.9	83.8
50–75	2.3	99.6	133.8	90.7
75–100	0.8	99.8	64.4	94.0
100 +	0.7	100.0	109.8	100.0
Total	**413.3**		**1,925.8**	

Source: U.S. Department of Agriculture, Soil and Water Resources Conservation Act, *Soil and Water Resources Conservation Act—1980 Appraisal Part II: Soil, Water and Related Resources in the United States: Analysis of Resource Trends* (Washington, D.C.: U.S. Government Printing Office, 1981), p. 59.

Of course, not all land erodes at the same rate. Although soil erosion occurs everywhere in the United States, the most serious problems tend to be concentrated in specific areas on a small proportion of the land. Almost one-third of the nation's agricultural land is experiencing very little erosion (less than 1 ton per acre per year), accounting for 2.6 percent of total erosion (figure 1.2). At the other extreme, slightly less than 3 percent of the agricultural land (including some of the best cropland) is eroding at a rate of more than 25 tons per acre per year—almost one-third of the nation's total erosion. And 6 percent of the erosion is occurring on the 0.2 percent of the land eroding at more than 100 tons per acre.[12]

Over half the cropland erosion occurs in the Corn Belt and the Northern Plains (figure 1.3). However, Hawaii and the U.S. Caribbean islands, as well as the Appalachian states, have higher erosion rates per acre of land. Because the Corn Belt is one of the most productive agricultural areas in the world, the erosion there is of particular concern. The region's intense cultivation in row crops (such as corn and soybeans) results in over one-third of the cropland eroding at an annual rate of 5 tons per acre or more.

Other regions of the United States also have serious local erosion problems. For instance, the combination of steep slopes and easily erodible soil in Washington, Oregon, and Idaho produces erosion rates of 50 to 100 tons per acre annually in some cropped areas.[13] And in some parts of Maine, the intense cultivation used in growing potatoes has resulted in a loss of up to 24 inches of topsoil.

These high rates of erosion can cause serious problems on the affected farmland. Five tons of soil erosion per acre per year is considered to be "tolerable." Yet, under natural conditions, it can take 300 to 1,000 years (particularly in arid areas) to form one inch of soil.[14] At the five-ton rate, it takes only 33 years to lose that inch. Even in many agricultural areas that are not particularly arid, U.S. cropland soil is being eroded significantly faster than it is being formed. And the soil being eroded is the topsoil—the soil that is most productive for farming because it usually has the highest organic content and most nutrients and is the best at holding moisture.

Relatively little information exists on the extent to which erosion decreases soil productivity.[15] Although average productivity nationally has continued to increase at a rapid rate, this growth likely has been due to the effects of erosion being more than offset by new crop varieties and increased use of fertilizer as well as other inputs such as tractors and water. Usually, a farmer can compensate for the loss of natural productivity by applying more fertilizer, at least until ero-

Figure 1.3
Cropland Erosion, by Region, 1977

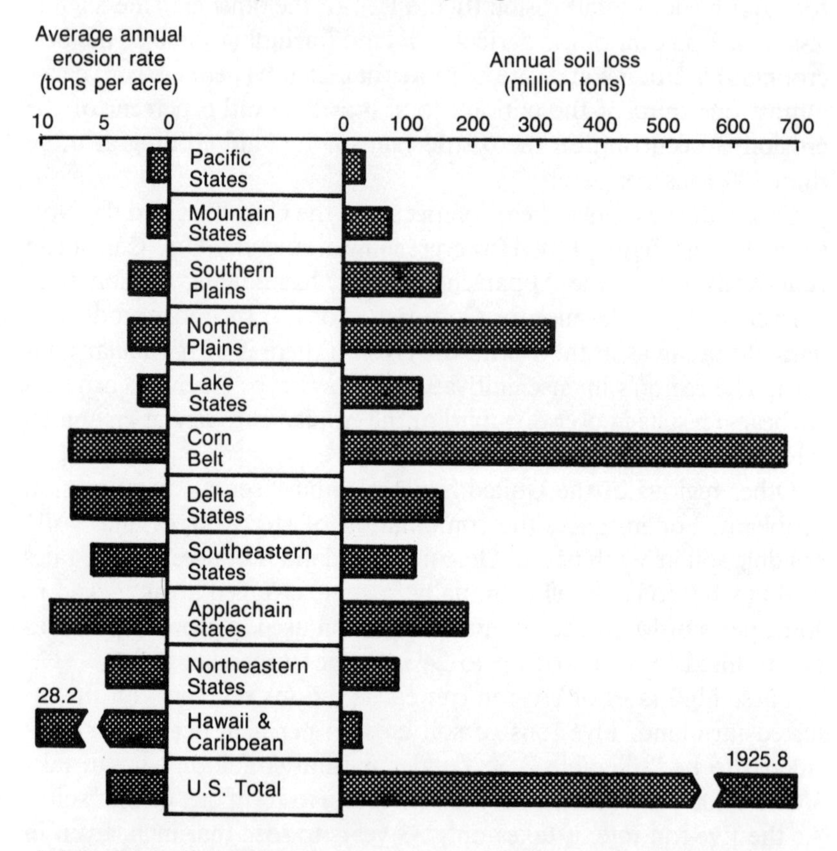

Includes sheet and rill erosion only

Regions are U.S. Department of Agriculture farm procduction regions.

Source: U.S. Department of Agriculture, Soil Conservation Service, *1980 Appraisal Part I: Soil, Water, and Related Resources in the United States: Status, Conditions, and Trends* (Washington, D.C.: U.S. Government Printing Office, 1981), fig. 24, p. 106, and fig. 25, p. 108.

sion becomes so severe that the soil's capacity to hold water is substantially affected. Still, at some point, productivity may drop suddenly and substantially—by as much as 40 to 60 percent.[16] Moreover, one automatic effect of farmers compensating for losses in natural productivity by using fertilizers is an increase in their costs of production.

Impacts off the Farm

The problems caused by soil erosion are not, however, restricted to reduced productivity and other impacts that occur on the farm. Indeed, the off-farm impacts of soil erosion may have more serious economic impacts than the on-farm impacts have.[17]

The awareness that soil erosion can have impacts off a farm is nothing new. Early American settlers found that the muddy waters caused by erosion filled millponds and river channels with silt, increased flooding, covered productive fields with mud, obliterated fish spawning beds, and turned clear deep rivers into shallow, muddy channels.[18] In this century, extensive soil erosion in Illinois has helped create an "ecological catastrophe," with lakes filled up, fish populations wiped out, and wildlife habitat destroyed.[19] Further west in Utah, deforestation and overgrazing had decimated the naturally clear streams of the Wasatch Mountains in Utah by the 1930s, causing serious problems of siltation and floods greater than any that had been experienced there "for at least twenty thousand years."[20]

Water-resource engineers have long had to take account of erosion-related sediment loads in designing canals and dams. Reservoirs have substantial portions of their capacity allocated to no other purpose than being filled in by mud. When such a "sediment pool" is insufficient to contain the volume of sediment entering a reservoir, the sediment begins to take up valuable storage capacity that would otherwise be used to supply drinking, industrial, or irrigation water or to prevent downstream floods. Eventually, the reservoir may be little more than a long, flat plain, with its former valley filled with choice topsoil.

Other, perhaps more serious problems arise because erosion from cropland includes not only natural soil but also fertilizers and pesticides, the use of which has increased very rapidly in recent decades. Thus, even if the problem of erosion itself has not worsened (and there is no solid evidence indicating whether this is the case), the off-farm impacts of soil erosion may have worsened because of the increased amounts of pollutants that are applied to—and washed off—farmland. Pesticides can be toxic to fish and, potentially, to humans. Fertilizers cause algae to grow, accelerating the eutrophication of lakes and reservoirs. As algae die and decay, they use up oxygen, sometimes reducing oxygen levels so much that some fish species can no longer survive. These off-site pollution problems are exacerbated when farmers try to compensate for lost productivity by using more fertilizers, pesticides, and other inputs, without reducing the amount of erosion.

These chemicals and sediments carried off cropland are one type of "nonpoint-source" water pollution—that is, pollution that is not discarded from a sewer, industrial discharge pipe, or other point source. Other significant nonpoint sources include storm-water runoff from urban streets, construction, and plant sites; mine drainage; runoff from forestland, rangeland, and pastureland; and leakage from septic tanks.

Agricultural nonpoint sources (including cropland, rangeland, and pastureland) are thought to be major contributors both of conventional pollutants, such as sediment, biochemical oxygen demand (BOD),* and dissolved solids, and of chemical pollutants, such as fertilizers and pesticides. In addition, rangeland and pastureland, along with feedlots (which are treated as point sources), can be serious sources of fecal coliform bacteria. (Nonagricultural sources, it should be noted, can contribute many of these pollutants, as well as significant amounts of heavy metals such as lead and cadmium and other substances classified as "toxic" under the Clean Water Act.)

Water-pollution-control officials have become more aware of these nonpoint-source pollution-control problems as the attempts to control discharges from point sources have begun to be effective. Nevertheless, few if any actual measurements of the impact of agricultural and other nonpoint-source pollution on water quality exist, and few efforts have been made to control them.

There have, however, been several efforts to estimate the relative impact of these sources by using models or other estimating techniques (figure 1.4).[22] According to these analyses, nonpoint sources contribute as much as 73 percent of total BOD loadings, 99 percent of suspended solids, 83 percent of dissolved solids, 88 percent of nitrogen, 84 percent of phosphorus, and 98 percent of bacteria loads in U.S. waterways. As industrial and municipal point sources implement the

---

*Technically, BOD is not a pollutant, although it is often referred to as one by water-quality experts. Rather, according to engineer Thomas R. Camp, it "is a measure of the concentration of decomposable organic matter in [a sample of sewage, industrial waste, or polluted water]. It has become the most useful single determination in the routine examination of sewages, industrial wastewaters and treatment plant effluents and, especially, in the examination of the receiving waters in pollution studies. The BOD concept involves not only the amount of organic material which is decomposable by bacteria but also the time rate at which it will decompose aerobically."[21]

Figure 1.4
Estimates of Pollutant Loadings from Nonpoint Sources

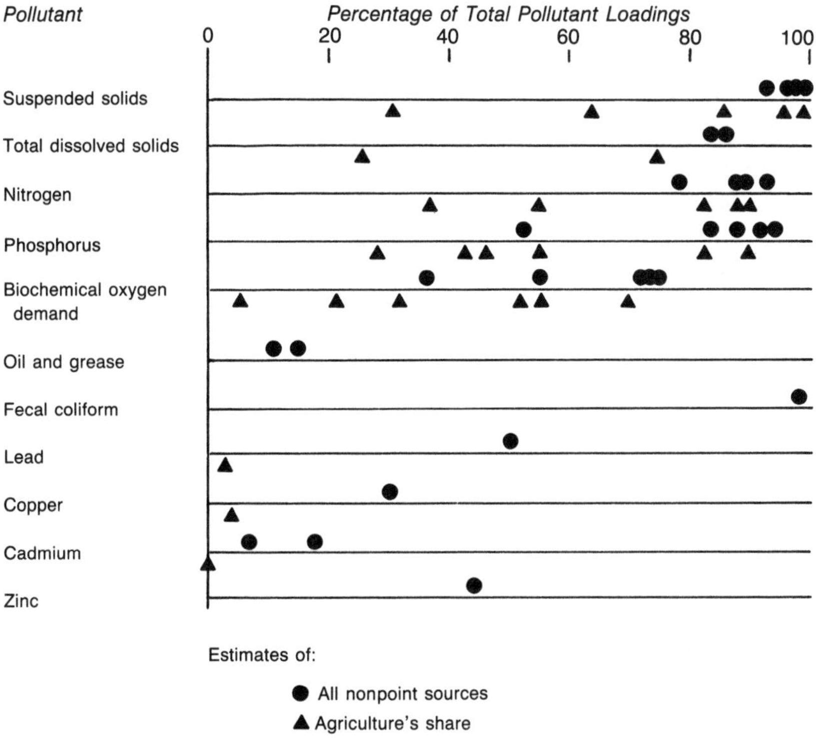

Pollutant

Percentage of Total Pollutant Loadings

0 20 40 60 80 100

Suspended solids
Total dissolved solids
Nitrogen
Phosphorus
Biochemical oxygen demand
Oil and grease
Fecal coliform
Lead
Copper
Cadmium
Zinc

Estimates of:

● All nonpoint sources

▲ Agriculture's share

Source: Edwin H. Clark II, "Estimated Effects of Nonpoint Source Pollution" (Research report, The Conservation Foundation, Washington, D.C., 1984).

pollution-abatement requirements imposed by the Clean Water Act, these percentages will increase.[23]

Much less information exists about the amount of toxic and other nonconventional pollutants that nonpoint sources contribute to U.S. waterways, but nonpoint sources can be major suppliers of these as well. For instance, in the Great Lakes they are thought to be substantially more important than municipal treatment plants as a source for heavy metals.[24] A report by the Great Lakes Commission concluded that urban-runoff control may be the most effective way to reduce metal loading for the lakes.[25] For Wisconsin and Michigan, rural and

Figure 1.5
Relative Importance of Nonpoint-Source Pollution

USGS Region	Number of accounting units for which causes of problems were given	Percentage of accounting units listing causes for which nonpoint sources were identified as contributing factors
1 New England	12	50
2 Mid-Atlantic	14	100
3 South Atlantic-Gulf	32	100
4 Great Lakes	18	89
5 Ohio	21	100
6 Tennessee	4	100
7 Upper Mississippi	11	100
8 Lower Mississippi	16	100
9 Souris-Red-Rainy	5	100
10 Missouri Basin	42	98
11 Arkansas-White-Red	21	100
12 Texas-Gulf	21	90
13 Rio Grande	11	91
14 Upper Colorado	7	100
15 Lower Colorado	7	86
16 Great Basin	7	100
17 Pacific Northwest	9	89
18 California	8	100
19 Alaska	3	0
20 Hawaii	5	100
21 Caribbean	2	50
Other (Guam, etc.)	4	100
Total U.S.	**280**	**94**

Source: U.S. Environmental Protection Agency, Office of Water and Waste Management, Monitoring and Data Branch, "Assessment of Ambient Conditions—Water Quality Portion, 1980" (Draft report); and U.S. Department of the Interior, Geological Survey, *National Water Summary 1983* (Washington, D.C.: U.S. Government Printing Office, 1984), p. 47.

Figure 1.5 (continued)
Relative Importance of Nonpoint-Source Pollution

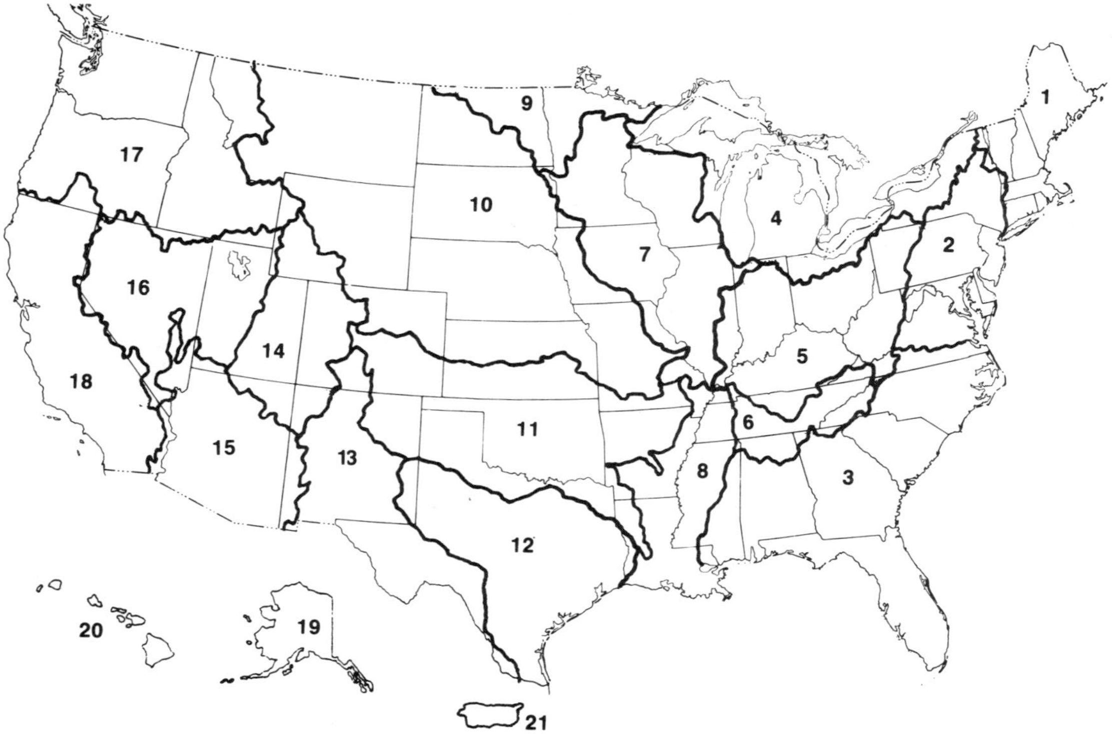

Source: U.S. Environmental Protection Agency, Office of Water and Waste Management, Monitoring and Data Branch, "Assessment of Ambient Condition—Water Quality Portion, 1980" (Draft Report); and U.S. Geological Survey,

urban runoff provide consistently higher heavy-metal loadings (except for copper in Lake Erie) than do the state's industrial sources.[26] Still, it is not clear to what extent these findings can be generalized.

These estimates pertain only to relative loading rates. It is difficult to assess the relative impact of nonpoint sources of pollution on water quality. The U.S. Environmental Protection Agency (EPA) does prepare biannual reports, pursuant to section 305(b) of the Clean Water Act, that summarize the best estimates of state water-pollution-control officials on the extent and causes of water-quality problems. A draft report of the 1980 survey, for instance, indicated that 94 percent of the accounting units for which causes of water-quality problems were given identified nonpoint sources as controlling factors (figure 1.5). In many cases, nonpoint sources were thought to contribute to only one or two specific water-quality problems that were identified, whereas point sources were considered to be more important in causing the other problems. Even in those cases in which a nonpoint source was identified as a contributor to a problem, it may have been only one of many causes. As a result, it is not possible from the information presented in the draft report to estimate the relative importance of nonpoint sources in causing pollution, although the report did indicate that the problem was quite extensive.[27] (Figure 1.6 provides a similar breakdown for the 1977 305(b) study.)

The 1982 section 305(b) report did not include a thorough assessment of water-quality problems in all the states. Nevertheless, it concluded that nonpoint sources were the most important cause of water degradation in almost one-fifth of the states and were a problem in almost all of them. Indiana and Illinois, for instance, reported that agricultural runoff was responsible for a significant percentage of fish kills.[28]

Another study by the U.S. Fish and Wildlife Service in association with EPA documented the impact of nonpoint-source pollution on the nation's aquatic wildlife. Nonpoint sources in general were identified as the probable source of water-quality problems in 367,244 stream miles across the United States, or 38.4 percent of the total stream miles surveyed.[29] The major cause was agricultural sources that, by themselves, were responsible for water-quality problems in 29.5 percent of the stream miles and were considered a "major concern" in 17.3 percent of the stream miles.[30] In addition to water-quality problems, the investigators identified degraded habitat as one of the major classes of problems—occurring in almost half the stream miles

Figure 1.6
Drainage Basins Affected by Nonpoint Sources of Pollution, by Type of Nonpoint Source, 1977[1]

Geographical region[2]	Number of basins	Percentage of basins affected				
		Urban runoff[3]	Agriculture	Mining	Silviculture	Construction
Northeast	40	70	55	20	10	15
Southeast	47	57	62	15	30	2
Great Lakes	41	54	59	41	15	7
North Central	35	54	89	40	6	6
South Central	30	50	87	53	13	0
Southwest	22	23	73	36	5	0
Northwest	22	23	55	23	27	23
Islands	9	67	78	0	0	67
Total	**246**	**52**	**68**	**30**	**15**	**9**

[1] Inlcudes those basins where some (or all) stream segments have a problem with a pollutant that is neither minor nor insignificant according to state officials.

[2] The regional drainage-basin groupings used here do not conform precisely to the 10 federal regions used in the discharge-trends analyses.

[3] Does not reflect drainage basins affected by combined sewer overflows.

Source: U.S. Environmental Protection Agency, *National Water Quality Inventroy: 1977 Report to Congress*, EPA 440/4–78–001 (Washington, D.C.: U.S. Environmental Protection Agency, 1978), p. 15; cited in U.S. Environmental Protection Agency, *Environmental Outlook: 1980* (Washington, D.C.; U.S. Government Printing Office, 1980).

surveyed—adversely affecting fish in the nation's waterways.[31]* And the most important cause of the degraded habitat in 27.9 percent of the stream miles surveyed was excessive siltation.[33]

In 1984, the Association of State and Interstate Water Pollution Control Administrators (ASIWPCA) also published an assessment of water-quality conditions in the United States.[34] For that report, ASIWPCA asked state officials to identify the primary causes preventing their streams from supporting their designated uses.[35] Nonpoint sources ranked first in 26 states and second in 13 others. Forty states reported that nonpoint sources need to be controlled if water quality is to continue to improve.[36] Agriculture was found to be the most significant nonpoint-source problem (figure 1.7).

Finally, nonpoint sources of pollution may have been the cause of

*The third, and quantitatively most important, major cause was insufficient water quantity, but in half of the adversely affected stream miles, the lack of adequate amounts of water was a natural condition.[32]

Figure 1.7
Severity and Extent of Nonpoint-Source Pollution Problems for 45 States, 1982[1]

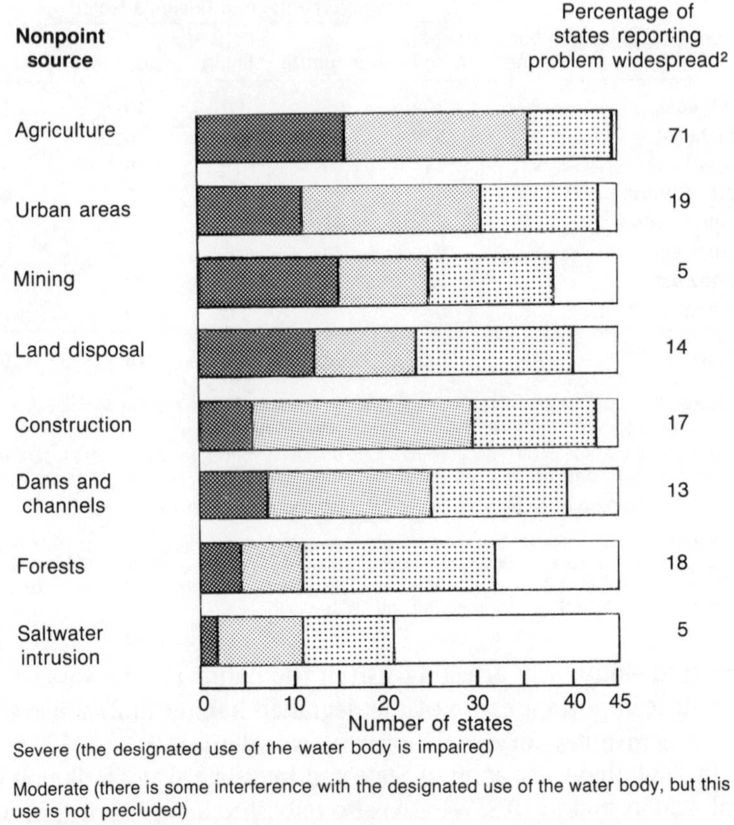

Severe (the designated use of the water body is impaired)

Moderate (there is some interference with the designated use of the water body, but this use is not precluded)

Minor (minimal effects on the designated use)

Not applicable or not reported

[1] Hawaii, Ohio, Oklahoma, Pennsylvania, and Texas did not provide any information on nonpoint sources of pollution.
[2] The nonpoint source is considered to cause a widespread problem if it affects 50 percent or more of a state's waters.

Source: Association of State and Interstate Water Pollution Control Administrators, *America's Clean Water*, prepared for U.S. Environmental Protection Agency (Washington, D.C.: Association of State and Interstate Water Pollution Control Administrators, 1984), p.11.

one of the more disturbing findings in the ASIWPCA study. The report found that, despite the billions of dollars spent abating point sources of pollution over the past decade, the water quality in lakes has gotten worse in more instances than it has gotten better. Many of the impacts of nonpoint sources of pollution would be expected to be more evident in lakes than in tributary streams (see chapter 2).

PURPOSE OF THIS STUDY

This study focuses on agricultural nonpoint-source pollution; specifically, the water pollution and other problems caused by sediment and other contaminants that are carried off farms following cropland erosion. It attempts to accomplish two goals. The first is to identify, document, and estimate the seriousness of the various off-farm impacts of soil erosion, including problems caused by contaminants associated with sediment as well as damages caused by sediment itself. The second is to assess policies that could be adopted to efficiently reduce those off-farm impacts.

In pursuing the first goal, the study does not, in most cases, distinguish the impacts resulting from the erosion of agricultural land from those impacts associated with erosion from all sources. As has been indicated, agricultural lands are only one (albeit the most important) source of erosion. Attributing certain impacts to certain sources is almost impossible since sediment from an upstream construction site looks the same—and causes many of the same problems—as sediment eroded from a nearby cornfield.*

Chapter 2 of this study summarizes the chemical, physical, hydrological, and ecological principles relevant to understanding the process by which eroded soil causes different types of off-farm impacts. Chapters 3 and 4 then identify the range of possible off-farm impacts and summarize an extensive literature search undertaken to document the extent to which those impacts occur and how serious they are. Chapter 5 attempts to estimate the economic costs of the impacts documented in chapters 3 and 4. Chapter 6 summarizes information on the costs and effectiveness of various techniques available for controlling nonpoint-source pollution from agricultural lands. Finally, chapter 7 focuses on some of the public policy issues involved in developing a program to control off-farm impacts, with a particular emphasis on targeting.

Much remains to be learned about the problem of off-farm impacts. The best information in this report is probably the information on the existence of the different types of impacts. The estimates on their magnitude and economic costs should only be considered indicative, at best. In particular, the reader should not make the error of assum-

*This study deals only with those impacts associated with erosion that ends up in water. This ultimately includes most of the erosion caused by wind, but the study does not deal with the impacts of the dust before it enters the water system.

ing that impacts for which magnitude and cost estimates have not been made are of less importance than the impacts for which such estimates exist. The fact that someone has not assigned a number to something does not make that something unimportant.

REFERENCES

1. Stanley W. Trimble, *Man-Induced Soil Erosion on the Southern Piedmont, 1700-1970,* (Soil Conservation Society of America, 1974), p. 23.

2. Ibid., pp. 22-23.

3. Ibid., p. 75.

4. Ibid., p. 55.

5. James Guilluly, Aaron C. Waters, and A. O. Woodford, *Principles of Geology,* 3d Ed. (San Francisco: W.H. Freeman and Co., 1968), p. 235.

6. Aloys Arthur Michel, *The Indus Rivers: A Study of the Effects of Partition* (New Haven, Conn.: Yale University Press, 1976), p. 30.

7. For analyses of the impact of soil erosion on ancient civilizations, see, generally, J. Donald Hughes in collaboration with J. V. Thirgood, "Deforestation, Erosion, and Forest Management in Ancient Greece and Rome," *Journal of Forest History* 26, no. 2 (1982):60-67; Fairfield Osborn, *Our Plundered Planet* (Boston: Little, Brown and Co., 1948); G. V. Jacks and R. O. Whyte, *The Rape of the Earth: A World Survey of Soil Erosion* (London: Faber and Faber, 1939); Hugh Hammond Bennett, *Soil Conservation* (New York: McGraw-Hill Book Co., 1939); J. H. Stallings, *Soil Conservation* (Englewood Cliffs, N.J.: Prentice-Hall, 1957); and R. P. Beasley, *Erosion and Sediment Pollution Control* (Ames, Iowa: Iowa State University Press, 1972).

8. Hughes with Thirgood, "Deforestation, Erosion, and Forest Management in Ancient Greece and Rome," p. 68.

9. Quoted in U.S. Department of Agriculture, *1980 Appraisal Part I: Soil, Water, and Related Resources in the United States—Status, Condition, and Trends* (Washington, D.C.: U.S. Government Printing Office, 1981), p. 9.

10. Ibid., p. 98.

11. U.S. Department of Agriculture, Office of Governmental and Public Affairs, *Fact Book of U.S. Agriculture* (Washington, D.C.: U.S. Government Printing Office, 1977), p. 37.

12. U.S. Department of Agriculture, *1980 Appraisal Part II: Soil, Water, and Related Resources in the U.S.—Analysis of Resource Trends* (Washington, D.C.: U.S. Government Printing Office, 1981), pp. 154-155.

13. U.S. Department of Agriculture, *1980 Appraisal Part I,* pp. 107, 109.

14. David Pimentel et al., "Land Degradation: Effects on Food and Energy Resources," *Science* 194 (October 8, 1976):150.

15. Pierre R. Crosson has published one of the more extensive assessments of these on-farm effects. See Crosson, *Productivity Effects of Cropland Erosion in the United States,* published for Resources for the Future (Baltimore: Johns Hopkins University Press, 1983).

16. R. Neil Sampson, *Farmland or Wasteland, A Time to Choose: Overcoming the Threat to America's Farm and Food Future* (Emmaus, Pa.: Rodale Press, 1981), p. 283.

17. Bartelt Eleveld and Harold G. Halcrow, "How Much Soil Conservation Is Optimum for Society?" in Harold G. Halcrow, Earl O. Heady, and Melvin L. Cotner, eds., *Soil Conservation Policies, Institutions, and Incentives,* prepared for North Central Research Committee III (Ankeny, Iowa: Soil Conservation Society of America, 1982), pp. 246-49; and M. T. Lee et al., *Economic Analysis of Erosion and Sedimentation: Hambaugh-Martin Watershed* (Urbana-Champaign, Ill.: University of Illinois at Urbana-Champaign, Illinois Agricultural Experiment Station, 1974), pp. 31-33.

18. Trimble, *Man-Induced Soil Erosion on the Southern Piedmont, 1700-1970,* p. 119 and elsewhere.

19. James Krohe, Jr., "The Illinois River: An Ecological Disaster,'' *Illinois Issues* 7, no. 2 (1981):16-17.

20. Jacks and Whyte, *The Rape of the Earth,* p. 32.

21. Thomas R. Camp, *Water and Its Impurities* (New York: Reinhold Publishing Corp., 1963, p. 243.

22. See Edwin H. Clark II, "Estimated Effects of Non-point Source Pollution" (Washington, D.C.: The Conservation Foundation, 1984) for a more extensive summary of these.

23. Ibid.

24. Thomas McHeidke, Douglas J. Scheflow, and William C. Sonzogni, "U.S. Heavy Metal Loadings to the Great Lakes: Estimates of Point and Nonpoint Contributions" (Great Lakes Environmental Planning Study, Contribution no. 12, 1980); and International Joint Commission, Great Lakes Science Advisory Board, *1980 Annual Report: A Perspective on the Problem of Hazardous Substances in the Great Lakes Basin Ecosystem* (Windsor, Ontario: International Joint Commission, 1980).

25. McHeidke, Scheflow, and Sonzogni, "U.S. Heavy Metal Loadings to the Great Lakes,'' p. 1.

26. Rose Ann C. Sullivan, Douglas J. Scheflow, and William C. Sonzogni, "The Relative Significance of U.S. Industrial Heavy Metal Loads to the Great Lakes" (Great Lakes Environmental Planning Study, Contribution no. 16, 1980), p. 3.

27. U.S. Environmental Protection Agency, Office of Water and Waste Management, Monitoring and Data Branch, "Assessment of Ambient Conditions—Water Quality Portion, 1980" (draft report).

28. U.S. Environmental Protection Agency, Office of Water Regulations and Standards, *National Water Quality Inventory: 1982 Report to Congress* (Washington, D.C.: U.S. Environmental Protection Agency, 1984).

29. Robert D. Judy et al., *1982 National Fisheries Survey,* vol. 1, *Technical Report: Initial Findings*, prepared for U.S. Department of the Interior, Fish and Wildlife Service, and U.S. Environmental Protection Agency, Office of Water, FWS/OBS-84/06 (Washington, D.C.: U.S. Department of the Interior, 1984), p. 28.

30. Ibid., p. 30.

31. Ibid., p. 22.

32. Ibid., p. 33.

33. Ibid., p. 39.

34. Association of State and Interstate Water Pollution Control Administrators, *America's Clean Water: The States' Evaluation of Progress, 1972-1982,* prepared in cooperation with the U.S. Environmental Protection Agency (Washington, D.C.:

Association of State and Interstate Water Pollution Control Administrators, 1984), pp. 10-11.

35. Ibid.

36. L. P. Gianessi and H. M. Peskin, "Analysis of National Water Pollution Control Policies: 2. Agricultural Sediment Control," *Water Resources Research* 17, no. 4 (1981):803.

2. The Basic Processes

The linkage between the erosion of a farmer's field and the various types of off-farm impacts that it can cause is neither direct nor simple. The soil itself is not a homogeneous substance. Physically, it is usually a mix of different particle sizes: clays (less than 0.002 millimeters in diameter), silts (0.002 to 0.05 millimeters), sands (0.05 to 2 millimeters), and sometimes gravels. The separate particles may be either highly consolidated, leaving very few voids, or loosely consolidated, with over 50 percent of the soil volume being air space.

Chemically, natural soil is a combination of mineral particles created by the weathering of rocks and organic substances, the latter resulting primarily from the decomposition of plant and animal residues. Soil also contains a number of different bacteria and other organisms. Human activities may have added additional chemical and biological components such as fertilizers and pesticides. All these factors—the chemical and biological constituents of the soil, as well as the size of the particles that compose it—influence how soil erodes and what types of problems it creates downstream.

THE PHYSICAL PROCESSES

Most erosion begins as raindrops hit the soil. An individual raindrop may be well over one-tenth of an inch in diameter and may fall at a velocity of approximately 25 feet per second.[1] Two inches of rain falling during a large thunderstorm is equivalent to over 225 tons of water landing on a single acre. The force of this rain breaks up soil granules and detaches individual particles—lifting them as much as 3 feet into the air and tossing them as much as 5 feet to the side. More than 100 tons of soil per acre may be detached by raindrops in a single rain.[2]

At the beginning of a storm, most of the rainfall seeps immediately into the soil. But the surface soil soon becomes saturated. Infiltration slows, and a muddy film of water on the surface catches the small soil particles thrown up by the raindrops. As this muddy water seeps into the ground, the small particles in it clog the soil surface, reducing the rate of infiltration even further. As a result, the film of water on the surface builds up and begins to move down the slope.

As the water moves over the surface of the soil, carrying along the suspended soil particles, it causes "sheet erosion." The expanding volume of water flowing over the surface has increasing erosive force, which is further heightened by already eroded soil particles bumping against and dislodging other particles still attached to the surface. In addition, the falling drops of rain continue to detach additional soil particles and, by agitating the water film, help keep those particles in suspension. If the rain stops, or an obstruction such as a tree prevents the raindrops from striking the ground, much of the suspended soil settles out.

Sheet erosion occurs uniformly over an entire soil surface and, as a result, often is undetected. However, it can pose a very serious problem since it removes a high proportion of both the lighter organic particles that are important contributors to the soil's fertility and the chemical contaminants that attach themselves to those particles.

As its flow increases, the water tends to concentrate in small channels called rills. The rills are often only a fraction of an inch deep and are rarely deep enough to interfere with a field's cultivation or other use. In fact, a farmer can easily obliterate them in the next tillage, often without even recognizing that they are present. Nevertheless, "rill erosion" can remove substantial amounts of soil by allowing water to move faster and thereby increasing its erosiveness.

In a few situations, however, these rills do enlarge into gullies that can destroy a field's usefulness. Once gullies begin to form, they can be difficult to control. They can enlarge rapidly, both in breadth as their banks are undercut and in length as the head of the gully cuts back into a field. "Gully erosion" is most likely to occur in areas where high rainfall intensity is combined with steeply sloped topography, where there are sharp drops at the edge of a field (perhaps caused by a stream running along the border) so that water cascades down as in a waterfall, or where the slopes are relatively soft, particularly if the soil is saturated with groundwater. Water seeping out of the ground along a slope can further promote the formation of gullies. In particularly erodible soils, gullies may develop from ruts caused by tractors being driven over soft ground.[3]

Providence Cave and the surrounding area in southwestern Georgia is a famous example of how significant gully erosion can become. Rain dripping off a barn roof supposedly began the erosion. Within 50 years, a ravine over 100 feet deep had covered an area of some 3,000 acres and had "engulfed a schoolhouse, two farm buildings, and much good farm land."[4]

Eventually, the water finds its way into an established channel—a brook, stream, or river—but the erosion can continue. The sides or bed of the waterway may erode, or the banks may be undercut and cave in, particularly during higher-than-normal flows.* This process, called "stream-bank erosion," is a normal and continuing process in many streams and rivers but may be exacerbated by the increased soil runoff that often accompanies more intense land use.

Determinants of Erosion

Five factors have the greatest impact on the amount of soil erosion that occurs from a field. These are expressed in a formula called the "universal soil-loss equation," which is commonly used for predicting erosion, particularly from agricultural lands.† It takes into account rainfall intensity and duration, soil erodibility, field topography, vegetative cover, and tillage practices. For noncropland, the vegetation and tillage variables are replaced by a land-use factor.[6]

Intensity and Duration of Precipitation

The intensity and duration of rainfall are both important in determining how much soil erosion occurs from a plot of ground. Snowfall and irrigation, and the runoff they cause, can also lead to erosion problems.

In general, the more intense the precipitation and the longer a storm lasts, the more erosion occurs. Rain showers providing less than one-half inch of precipitation are usually not important unless the ground is already saturated from another storm that occurred less than six hours earlier or unless at least a quarter inch of rain falls during a 15-minute period.[7]

The duration of a storm is important since, as indicated earlier, significant erosion begins to occur only after a soil surface has become saturated and a water film has begun to collect. During short storms, although the raindrops may dislodge soil particles, most of the water

---

*As is pointed out later in this chapter, if the water is carrying substantial amounts of soil with it, the result can also be just the opposite—heavy sedimentation in the river channel.

†The actual equation is $A = (R)(K)(L)(S)(C)(P)$, where A is the computed soil loss per unit area, R a rainfall and runoff factor, K a soil-erodibility factor, L a slope-length factor, S a slope-steepness factor, C a cover and management factor, and P a support-practice factor.[5] In this text, discussion of the L and S factors is combined.

seeps directly into the ground, depositing the soil particles back onto the surface. In the absence of surface runoff, very little soil will be carried off a field.

Rainfall/runoff erosion rates vary widely across the United States (figure 2.1). If all other factors are the same, for instance, the Gulf

Figure 2.1
Rainfall Erosion Intensity Isocurves

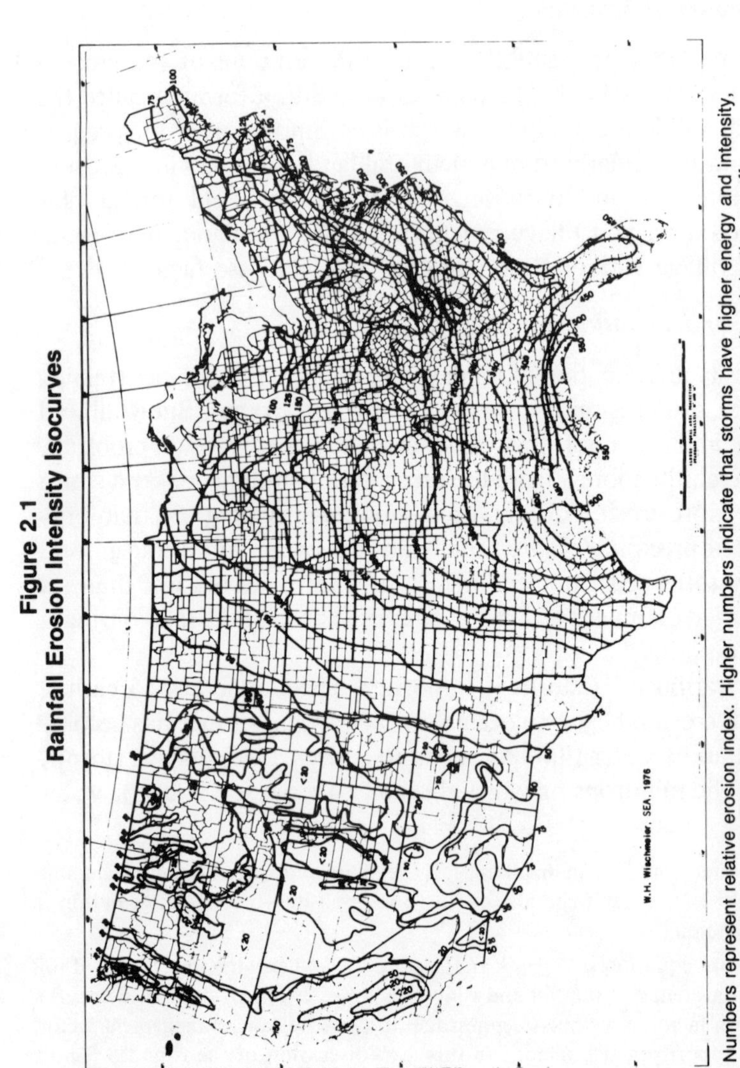

Numbers represent relative erosion index. Higher numbers indicate that storms have higher energy and intensity, thereby increasing their potential to cause erosion. Numbers do not include snowmelt or irrigation runoff.

Source: Walter H. Wischmeier and Dwight D. Smith, *Predicting Rainfall Erosion Losses: A Guide to Conservation Planning*, prepared for U.S. Department of Agriculture, Science and Education Administration, Agriculture Handbook 537 (Washington, D.C.: U.S. Government Printing Office, 1978), fig. 1.

Coast normally has erosion rates 25 times higher than those in the West. Moreover, in many areas of the country, much of the total erosion occurring during a year takes place during only one or two major spring storms. However, in other areas, more moderate storms also add significant amounts of erosion.[8]

Soil Erodibility

Different types of soil have different propensities to erode. Gravel does not erode easily, but then neither do sticky clay soils. And characteristics such as chemical and mineralogical composition, degree of compaction, and soil moisture content can strongly influence a soil's erodibility.

Many of the characteristics that create a good agricultural soil—easy workability, high organic content, good water retention—are, ironically, also the characteristics that increase its erodibility. The most erosive soils are generally those composed primarily of small, loosely bound soil particles. These soils typically are easily crumbled by hand. Loess soils (loosely consolidated fine-grained soils deposited by the wind) are particularly good examples, though any soil with a high proportion of silt tends to be easily eroded.

Soils high in organic content (decomposed plant and animal matter) also tend to erode easily. Not only are the organic materials themselves light and therefore easily carried off, but they also tend to prevent the other soil components from sticking together strongly.

Finally, the soil's permeability influences its erodibility. Highly permeable soils (for instance, those having a high sand content) that allow rainfall to seep in quickly and allow relatively little runoff, tend to erode less than soils with low permeability. Sandy or gravelly soils have one-seventh the erodibility of such soils as silty loams.[9]

Field Topography

A field's length and slope interact in complicated ways to influence the amount of erosion that occurs (figure 2.2). In general, the longer the field and the steeper its slope, the more erosion occurs.

A longer field allows more water to accumulate as runoff (or overland flow), thereby increasing its erosion potential. The most severe erosion typically occurs at the bottom of a field, where the most runoff water has accumulated. The effect of field length on the average erosion rate is not linear, however. For instance, doubling the length of a field with a 5 percent slope from 200 to 400 yards increases the average erosion rate only 41 percent.[10]

A field's slope has a more significant impact than length on the

Figure 2.2
Relationship of Erosion to Length and Slope of Fields

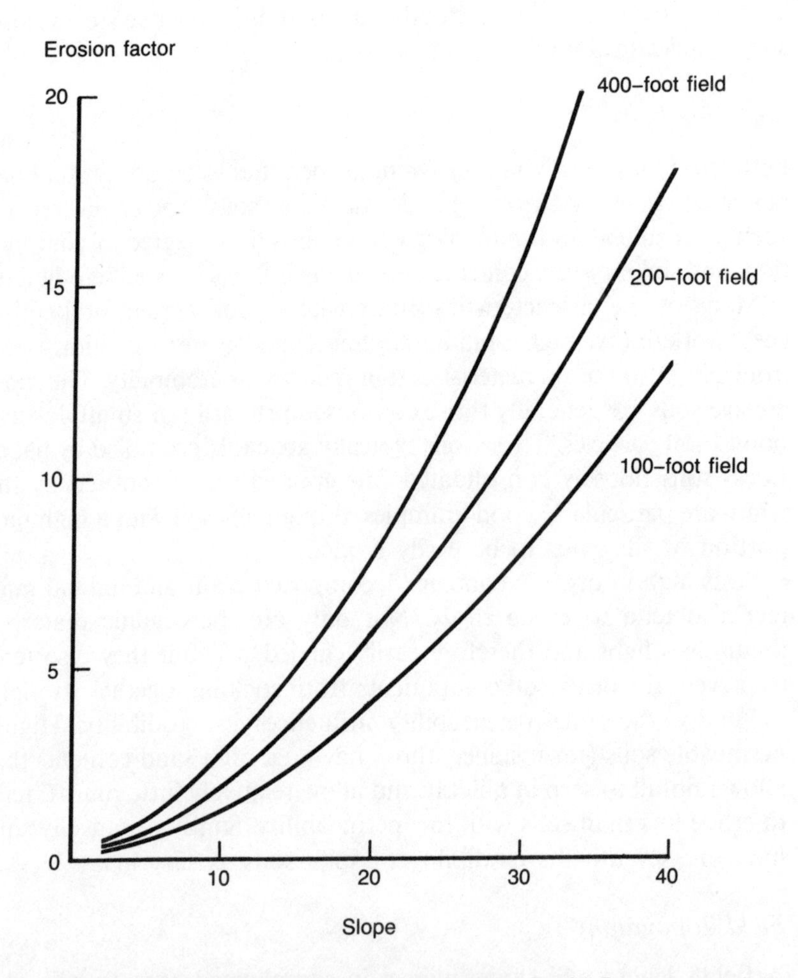

Source: Walter H. Wischmeier and Dwight D. Smith, *Predicting Rainfall Erosion Losses: A Guide to Conservation Planning*, prepared for U.S. Department of Agriculture, Science and Education Administration, Agriculture Handbook 537 (Washington, D.C.: U.S. Government Printing Office, 1978), p.13.

amount of erosion that occurs, since a steeper slope increases both the amount of water that runs off a field and the velocity at which it moves. Because, as has been explained, faster moving water can erode and carry more sediment, the amount of soil loss increases much more rapidly than the amount of water runoff.[11] A 200-foot field with a 35 percent slope will have 10 times the erosion of a field of the same size but with a 9 percent slope, or 100 times as much erosion as the same field with a 1 percent slope.*

Vegetative Cover

Soil cover can affect the amount of erosion that occurs in three ways. The first, the canopy effect, results when vegetation or other soil cover protects the soil surface from the impact of falling rain drops. The second is the surface-cover effect, which slows the water as it flows over the surface, reducing the water's sediment-carrying capability. The third is the infiltration effect, whereby certain types of cover can increase the amount of infiltration (water absorption by the soil) that occurs, thereby reducing surface runoff.

If a complete canopy of leaves (from trees, bushes, grasses, or structures) covers the soil, falling rain does not hit the soil surface with full velocity and consequently has less energy to detach soil particles and keep them stirred up. However, the canopy's ability to prevent erosion is inversely related to its height above the soil surface. High trees will block rainfall, but drops falling from the trees' leaves may gain substantial velocity again before they hit the ground (though this velocity is rarely as high as that attained by free-falling raindrops).[12] A canopy covering 80 percent of the surface at a height of 4 meters (about 13 feet) above the ground will by itself reduce erosion by 20 percent.[13] By contrast, the same canopy only half a meter (1.6 feet) above the ground will cause a 65 percent reduction in erosion.

Perhaps even more important is the percentage of the soil's surface covered by residual mulches or by close-growing vegetation. Such coverings not only interfere with raindrops striking the surface, thus providing a canopy effect; they also slow the rate at which the water flows over the surface, reducing its soil-carrying capacity. A full covering of mulch will reduce erosion by over 95 percent, regardless of the amount of canopy that lies above.[14]

Such groundcover can also increase the amount of infiltration and, therefore, reduce the amount of water flowing over the surface of the soil. This increased infiltration results from the reduced velocity with which the water flows over the surface and from the tendency for plant roots to break up the soil. Because of the multiple effects of vegeta-

---

*It is not, however, only the overall velocity of the water that is important. The water's turbulence can also affect the amount of erosion that occurs. A turbulent rivulet moving at moderate velocity may cause more erosion than one flowing faster but more smoothly.

tion, a partial vegetative cover is more effective at reducing erosion than the same percentage of mulch cover.*

The rapid shift of agricultural lands into row crops that occurred during the 1970s is thought to be a major reason for the current high rates of soil erosion. Land growing soybeans has been found to erode more than land growing corn.[15] And, in general, row crops such as soybeans, corn, and cotton tend to result in significantly more erosion than field crops such as hay and alfalfa. Row crops have both a lower percentage of canopy (particularly during the early growing season) and a higher percentage of bare, unprotected soil surface than field crops have.

Tillage Practices

The tillage factor in the universal soil-loss equation takes account of the use of farming techniques such as contour plowing, strip-cropping, or terracing that tend to lessen erosion. Because such practices either create furrows that run perpendicular to a field's slope or leave crop residues on the soil's surface, they reduce the rate at which water flows over the soil's surface and increase the amount of infiltration that occurs. By contrast, tillage practices that result in a smooth, clean field with no residues or vegetation left on its surface or, worse, that leave furrows running down its slope encourage high rates of soil erosion.

Sediment Delivery to a Waterway

Not all of the soil that erodes from a field ends up in a waterway. More than 90 percent of it may be deposited somewhere along the way, although this deposition may be only temporary.[16] Some is deposited in low spots or in places where the slope becomes flatter. Some of it is deposited along the edges of the field, especially if they are covered with vegetation. Some settles to the bottom of channels that carry water away from the field.

The difference between the amount of erosion estimated to reach the edges of a field and the amount of sediment that finally enters a waterway is expressed in what is known as a sediment-delivery ratio. Such a ratio can provide valuable information relative to the off-farm impacts of soil erosion. For instance, it can indicate how much of the soil that is eroded actually leaves a watershed to potentially cause impacts downstream. It also can show the proportion of the material

*However, a complete covering of mulch is almost as effective as a complete covering of vegetation.

that does not leave, thereby possibly causing impacts within the watershed.

Unfortunately, however, existing techniques for devising a sediment-delivery ratio are cruder and less accurate than are those for the universal soil-loss equation. One observer has concluded that "a characteristic relationship of sediment yields to erosion alone apparently does not exist,"[17] and another has stated that "the relationship between quantities eroded from the land surface and quantities delivered to some distant downstream location is exceedingly tenuous."[18]

Several factors can influence a sediment-delivery ratio:

Time Scale

A particle of soil's journey from a farm field to a waterway may be very leisurely, with several storms—and repeated erosion and deposition—required to move the soil off the field. Once the soil finally has left the field, several years of more storms may be necessary for it actually to reach a stream. Thus, a short-term sediment-delivery ratio may be quite low, although, in the long term, the ratio for soil that eventually reaches a waterway will be significantly higher.*

Relative Location

Not surprisingly, a higher proportion of soil eroded adjacent to a stream than of soil eroded distant from it ever gets to a waterway. To deal with this fact, an estimation technique referred to as the partial-area theory has been developed. This technique attempts to exclude from the sediment-delivery estimates all areas that are so remote from drainage channels that they contribute very little runoff and sediment to the waterways.[19]

Channel Density

The greater the number of channels per unit area of a watershed, the greater the speed and ease with which eroded materials reach a waterway. Typically, if the channel density of an area is low, a low propor-

---

*Of course, if the soil is redeposited on the land, it is counted again as newly eroded soil when it is picked up once more by a storm. So to argue that the long-term sediment-delivery ratio is higher than the short-term ratio also implies that the amount of new erosion associated with any storm is less than the total amount of erosion predicted by the universal soil-loss equation. Apparently, no studies exist that would allow the estimation of either of these factors.

tion of the sediment will reach a stream unless most of the erosion is occurring in or adjacent to gullies directly connected to the few channels that exist.

Watershed Size

Most hydrologists believe that the proportion of sediment transported out of a watershed usually will be lower for larger watersheds than for smaller ones. Much of the sediment tends to be deposited along the streams draining a watershed area.[20] In one study that used data from Texas, Oklahoma, the Missouri River Basin, and the southeastern Piedmont area, watersheds with drainage areas of 1 square kilometer were found to lose about 40 percent of their eroded soil, while the losses for watersheds with drainage areas of 100 square kilometers were put at only 10 percent (figure 2.3).

Figure 2.3
Sediment-Delivery Ratios and Watershed Areas

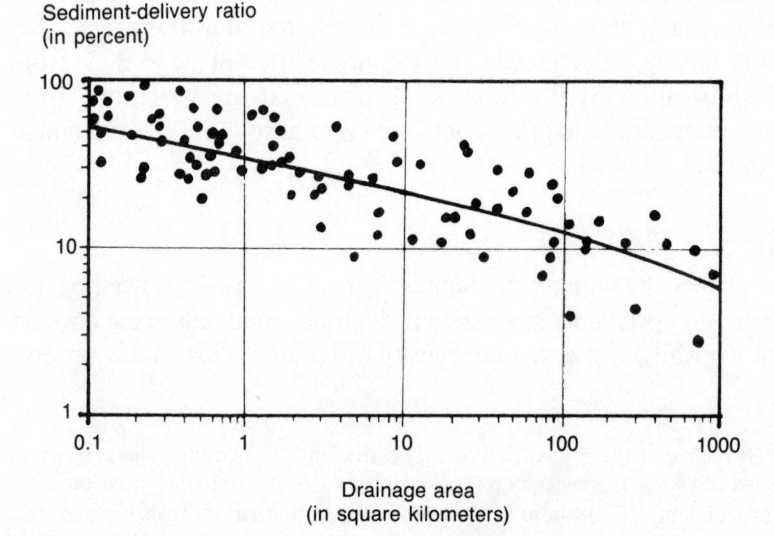

Source: Vladimir Novotny and Gordon Chesters, *Handbook of Nonpoint Pollution Sources and Management* (New York: Van Nostrand Reinhold, 1981), fig. 5-14, p. 190.

Soil Characteristics

Another important element in determining sediment delivery is the makeup of the eroded soil particles. The sediment-delivery ratio in a watershed with sandy soil, for instance, normally is low because sand particles rarely are transported far when they are eroded. By contrast, a watershed composed of predominantly clay soil usually has a high ratio, since clays typically remain suspended for long periods, even if the amount and velocity of runoff is relatively low. One study, reflecting this influence of soil type, used sediment-delivery ratios of "71-95 percent for clay, 61-84 percent for silty clay, 55 percent for silt, 35-51 percent for loam, and 20-30 percent for sand."[21]

Sediment is said to become "enriched" as it flows to a waterway—meaning that transported sediment contains a higher proportion of clays and other fine particles than does the soil from which it originates. This occurs not only because fine particles of soil are more easily eroded and carried than coarse ones but also because larger particles are more likely to be redeposited once they are disturbed.[22] The findings of an analysis of Mississippi cotton fields illustrate this enrichment process. The fields' soil was a fine loam, composed of approximately 53 percent sand, 23 percent silt, and 24 percent clay.[23] Yet virtually no sand particles were eroded from the fields. Instead, "about 83 percent of the sediment outflow was clay particles . . . and nearly all of the outflow, 98 percent, was [silt or clay]."[24]

The primary characteristics determining how fast a type of soil settles are density, size, and shape. Density is determined primarily by mineral composition. Gypsum particles and some clay minerals are only twice as heavy as pure water, while most silt is made out of quartz or feldspar, which is 2.6 to 2.7 times as heavy as water.[25] The higher the density of a soil particle, the faster it settles. Similarly, a large particle settles faster than a small one of the same density, and a spherical particle settles faster than a flat one.

Other Factors

As should be expected, some of the factors that influence the universal soil-loss equation—particularly, rainfall intensity and duration, and field topography—also affect sediment-delivery ratios. The increased runoff that results from a more intense, longer storm carries with it a more-than-proportional increase in sediment.[26] In addition, a flat area generally delivers less sediment to a stream than does a steep watershed with relatively well-defined channels.

Sediment Transport within a Waterway

Once sediment has eroded from a field and reached a waterway, probably the most important factor determining the magnitude and location of downstream impacts is how quickly it is redeposited. In many respects, the factors that influence sediment transport are similar to those that determine the amount of sediment initially carried off a field. For instance, the characteristics of the sediment itself continue to be important in determining whether the soil particles settle in or along the waterway or whether they continue to be transported.

However, after the sediment enters a waterway, an additional set of factors—the characteristics of the waterway itself—become increasingly important in the transport process. The turbulence and viscosity of the water in which soil particles are suspended help determine the sediment's settling rate. In turbulent water, the particles are being continually resuspended rather than allowed to settle to the bottom. Similarly, as water becomes more viscous, sediment settles more slowly.

Types of Sediment Load

A river's sediment load is divided into two parts. One is "bed load," the material that is contained in the top layer of the river's bed and is continually, though slowly, being carried downstream. The second, "suspended load," is the sediment suspended in and mixed with the water.

Bed load moves along on the surface of a riverbed in a thin layer of particles, sometimes no thicker than a couple of sand grains. Individual particles are repeatedly picked up and redeposited, skipping along the bottom. As they land on the bed, they may nudge the next particle along. As bed load moves downstream, constant bumping and sliding wears the particles down. For this reason, and because rivers typically become less steep as they approach their mouths, the size of bed-load material tends to decrease as a river flows from its headwaters to its mouth.[27]

The amount of sediment transported as bed load depends principally on the size and density of the particles forming the river channel and on the slope and depth of the river. The particle characteristics determine how easily a river's flow can move the bed load. Generally, a larger, heavier particle moves more slowly; also, a round particle moves more easily than an angular one. A river's slope and depth determine how fast the water in it flows. Higher slopes, greater depths, and the resultant higher velocities increase bed-load movement.

Two other factors—river temperature and suspended-load levels—

can exert lesser influences on bed-load transport. The higher viscosity of cold water can create twice as much bed-load transport as would exist under the same flow conditions in warm water.[28] High suspended-load levels also increase water viscosity.

Bed load often moves very slowly. Much of it, particularly large particles, may not move at all except during floods. In some cases—for instance, when the slope of a river flattens sharply as it leaves a mountainous area—large rocks carried down from a waterway's upper reaches may remain in place until they wear down or there is an unusually large flood.

Suspended load moves downstream more rapidly. As with bed load, the amount of suspended sediment carried by a river increases with the quantity, depth, and, most important, the velocity of water in the river.[29] Of course, the same storms that increase a waterway's volume and speed also cause the most erosion from the land. Thus, in many cases, as a river's capacity to transport sediment increases, the amount of sediment supplied to the river also goes up.

During floods, both the concentration and quantity of sediment in a river reach their highest levels. However, concentration rises the most rapidly, since sediment quantity increases faster than water quantity. On many rivers, it is not unusual for more sediment to be transported during a single flood than during the remainder of a year. And, in extreme cases, over 50 percent (by weight) of a river's flood flow may be composed of suspended materials, though the percentage usually is much lower.[30] The maximum sediment concentration ever measured by the Tennessee Valley Authority showed only 8.5 percent sediment, and that was on a small creek draining an area laid bare by mining.[31]

Suspended-load concentration is determined largely by the presence in a river of very small soil particles (or "fines"). The more fines there are, the more materials the river will carry and the more uniformly the sediment will be distributed throughout the river.[32] Not only can fine particles be kept suspended in water more easily than coarse particles can; the fine particles also increase water's ability to pick up coarse particles from the riverbed and to keep them in suspension.[33] (Like bed-load transport, suspended-load transport is additionally influenced somewhat by water temperature; a river can carry higher suspended loads in cold weather than in warm weather.[34])

Standard engineering formulas used to predict suspended-load levels traditionally have related the concentration of a particular particle size in a waterway to its concentration in the riverbed. However, there is increasing evidence that much of the suspended load may not come from the bed at all.[35] Instead, in small watersheds, between 90 and

95 percent of the suspended sediment may be "wash-load"—material that has been carried directly off the land surface without being deposited on a stream bed.[36]

If an incoming sediment load is composed of substantial amounts of coarse material, much of the sediment will be deposited in the river channel, causing an aggradation (or rise) in the level of the riverbed. For instance, several tributaries of the Trinity River in Texas receive large quantities of sandy sediment as they flow through the West Cross Timbers Area in the northeastern portion of the state. Much of this sand settles in the streams' channels, raising their beds several feet and causing increased flooding and the creation of swamps.[37]

River-Channel Structure

A typical river's ability to transport sediment undergoes often substantial changes along the waterway's length. Its headwaters are in hills or mountains, where they drain relatively little area but have relatively steep channels. As the river reaches plains land, its slope flattens substantially. This is usually the most dramatic, but not the only, change in slope along the river's length. Indeed, there tends to be a continuous flattening of the slope as the waterway continues downstream. (Figure 2.4 shows this process for the Arkansas River.) At the same time, the

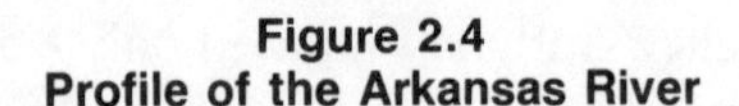

Figure 2.4
Profile of the Arkansas River

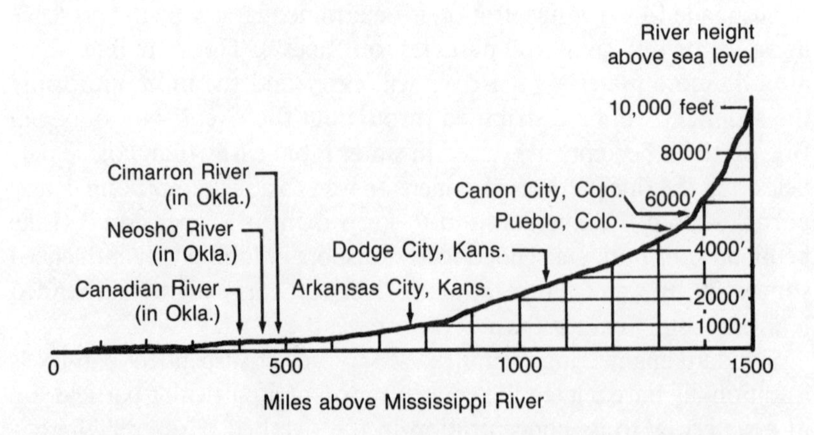

Source: James Gilluly, Aaron C. Waters, and A. O. Woodford, *Principles of Geology,* 3rd ed. (San Francisco: W. H. Freeman, 1968), fig. 12–14, p. 221.

drainage area, and therefore the volume of water in the river, constantly increases.

At any point during this trip, the shape, slope, and dimensions of the river's channel may be changed by the complicated interaction of several different factors. The most important of these are the amount of water in the river, the composition of the riverbed, geological or human interferences (for instance, rock outcrops or concrete embankments), and the sediment load that the river is carrying.

Generally, the more water a river carries, the larger its channel is. But most rivers carry substantially different amounts of water at different times of the year and from one year to another. In arid regions such as the West, a river channel may carry no water at all much of the time but periodically experience serious flooding. Both the high flows and average flows of a river significantly affect its form and its sediment-transport characteristics.

The composition of a river's bed is strongly interrelated with the river's sediment load. Together they help determine both the dimensions of a channel and the path of the river as it travels downstream (for instance, whether it is relatively straight or meanders back and forth). After reaching the plains, most river channels cut through alluvial materials that the river itself has deposited, either relatively recently or during previous geological periods. If the river's sediment is predominantly composed of gravel, the riverbed will be gravel as well. If the river is carrying mostly fine-grained material, the bed will be composed of the same.

Maintaining Equilibrium

A waterway may have to make significant adjustments to maintain balance among all the factors affecting sediment load and river dynamics. Indeed, even after an original problem has been solved, a river may have to continue its adjustment process for many years.[38] Predicting exactly what alteration will take place when a sediment load changes is very difficult. This is particularly true since changes in sediment load and in the quantity of water the river is carrying usually are directly associated, with an increase in one meaning an increase in the other. The eventual adjustment will depend on the relative increases in these two factors, as well as on the characteristics of the sediment itself and the river-channel structure.

The equilibrium of a river is rarely static. In an alluvial plain, for instance, a river continually erodes material from some locations and deposits it in others, affecting the waterway's dimensions. However, if the river's sediment load (or any of the other factors listed above) changes substantially, whatever equilibrium exists will be disrupted. The river may undergo substantial, and sometimes swift, adjustment in the location and form of its channel as it seeks a new balance. For instance, for decades it has seemed that the lower Mississippi River might shift its channel 70 miles west of New Orleans into the Atchafalaya floodway, leaving both New Orleans and Baton Rouge stranded on a quiet stream. The U.S. Army Corps of Engineers has had to construct massive control works to prevent this diversion from occurring.[39] A recent exhaustive study of problems along the lower Mississippi concluded that the "common thread to all changes in the basin is sediment."[40]

The most frequent cause of alteration in a waterway's sediment load or water volume is land-use change in the river's watershed. The clear-cutting of forests, a shift to more intensive row cropping, or increased urban development all can cause increased sediment loads and water runoff. This impact has been demonstrated untold times in the nation's history. A clear, quiet stream will be changed into a muddy river as its watershed is converted from forest to cropland, only to change back to a clear stream as the cropland is abandoned and the forest allowed to return.

The construction of a dam along a waterway also can significantly affect a river's equilibrium. For instance, as water approaches a reservoir, it begins to slow down. This causes the larger sediment particles it is carrying to settle out, while the finer particles are carried further into the reservoir before they settle. Some of the finest particles never settle, even in the largest reservoirs. Instead, they are eventually carried over the reservoir's spillway and continue moving downstream. (The extent to which U.S. reservoirs have been filled with sediment is discussed in chapter 3.)

The impact of a reservoir on a stream's sediment load depends on both the reservoir's size (relative to the amount of water flowing through it) and the characteristics of the sediment load. Relatively large reservoirs have a high "retention time"—that is, on the average, it takes a significant period of time (sometimes years) before water entering the upstream end of the reservoir flows over its spillway.* Small reservoirs may have a retention time measured in hours.

---

*The average retention time is computed by dividing the average annual inflow to the reservoir by its capacity. Thus, a reservoir having a storage capacity of 500,000 acre-feet, with an average annual inflow of 2 million acre-feet, has an average retention time of three months.

The longer a reservoir's retention time, the more opportunity there is for the suspended sediment to settle out of the water. Thus, a high proportion of the sediment load entering a relatively large reservoir will be deposited before the water carrying it leaves the reservoir. Such "trap efficiencies" for large reservoirs frequently exceed 95 percent (that is, more than 95 percent of the incoming sediment is deposited in the reservoir). For smaller reservoirs, the retention time and thus trap efficiency is usually lower—typically less than 70 percent.[41]

Yet ponds and small upstream reservoirs typically "silt up" at a significantly faster rate than do larger downstream reservoirs.[42] Three reasons explain this apparent anomaly. For one, since they are located in smaller watersheds, a higher proportion of whatever soil has been eroded in the watershed reaches the pond or reservoir. A second is that a reservoir's trap efficiency decreases less rapidly than does its volume. Therefore, even if the water flowing into a small reservoir had the same concentration of sediment as water flowing into a larger reservoir, the small reservoir would fill faster because the trap efficiency might be reduced by only 20 percent, though the volume might be reduced by 90 percent. Finally, upstream reservoirs often collect much of the sediment that would otherwise be deposited in a downstream reservoir. A study of reservoir sedimentation along the Trinity River in Texas concluded that the waterway's upstream reservoirs reduce the sediment load in a downstream portion of the river by 35 percent.[43]*

A new reservoir also can cause significant erosion to occur downstream of its dam. Because relatively silt-free water flows over the spillway, the river is forced to erode its bed and banks in an attempt to regain an equilibrium between its sediment load and channel structure. That erosion, in turn, causes the below-dam slope of the river to be flatter than it was before the dam was built. When a new equilibrium is achieved, possibly after decades, the river's sediment load will remain below its prereservoir level—probably because of the river's flatter slope and reduced water flow.[44]

CHEMICAL AND BIOLOGICAL FACTORS

The sediment eroded from fields rarely is composed solely of inert particles. Even when land has been untouched by humans, it natural-

---

*The upstream reservoirs, however, did not catch all the sediment that would otherwise have gone downstream. If they had, almost half the sediment load in the downstream portion of the river would have been reduced, since 49 percent of the watershed for that downstream portion was above the upstream reservoirs.

ly contains organic and other chemically or biologically active particles that can cause serious water-quality problems. When eroded and carried into a waterway, these components, along with plant residues and manure, can help cause excessive concentrations of nutrients in the water and can reduce its levels of dissolved oxygen.*

Human contact with land increases the proportion of chemically active particles in sediment. Modern farming practices, for instance, add significant amounts of chemical compounds to the fields—notably, chemical fertilizers and pesticides. Animal wastes on fields and pastures and in feedlots contribute still more bacteria and nutrients, and, if wastewater-treatment sludge is applied as a fertilizer, heavy metals and other contaminants can be added to the soil.

These chemically and biologically active compounds are, like the soil with which they are associated, subject to the effects of rainfall, runoff, and sedimentation. The same physical factors that transport eroded soil can carry such compounds from the fields and into the nation's waters. As the compounds flow in the runoff or river, they may be suspended as distinct particles, chemically bound to sediment particles, or dissolved. Once in a waterway, however, these contaminants can cause impacts different from those caused by sediment transport and deposition. These reactions occur because the contaminants, unlike soil particles, react chemically and biologically with their environment (for example, by making algae grow) and because the contaminants' own basic characteristics are changed in the waterway.

These factors make the transport and fate of contaminants inherently difficult to measure, and, as a result, much remains to be learned about this phenonmenon. Moreover, because the chemical specifications of many of the products of contaminants after they break down simply are not known, it is much more difficult to generalize about the overall effects of contaminants than about sediment itself.

Nutrients

Nutrients—in particular, nitrogen and phosphorus—are arguably the most significant contaminants in eroded soil and runoff. Their export from agricultural watersheds is influenced by all the factors that influence the erosion process itself, as well as by several others. Particularly important are soil type, fertilizer type and amount, farm-management practices, cropping patterns, animal presence and manure-storage practices, and weather.[45]

---

*This process, known as eutrophication, is explained in chapter 4.

Types of Nutrients

Some nutrient transport is entirely natural. Erosion may occur in soils that are naturally high in phosphorus or easily eroded organic matter containing nitrogen and phosphorus. Leaf litter falling into an undisturbed stream can provide a majority of the nutrients in a channel.

Still, most nutrient export from agricultural lands results from human activities. Although crop residues and feedlot wastes are also considered important, fertilizer application is the main source of agricultural nutrients.[46] In 1981, farmers used 12.3 million tons of nitrogen and 5.7 million tons of phosphorus, an increase of 323 percent for nitrogen and 110 percent for phosphorus since 1960.[47] When fertilizer is spread over land, only a small percentage of phosphorus is taken up by the first crop grown after the fertilizer is applied;[48] the proportion for nitrogen is higher.[49] Crop nutrient uptake is highest during periods of rapid growth.

Whatever fertilizer is not used by crops either chemically binds with the soil (and, therefore, if the soil is not eroded, may be available for subsequent crops), remains unbound in the soil, is dissolved in soil water, or is suspended or dissolved in irrigation or stormwater runoff. If the fertilizer is dissolved in soil water, it can leach into the ground. The fertilizer is particularly likely to be carried off the land if it is applied too early, when it is often carried off by snowmelt or early rains, or when it is applied just before a heavy storm. At other times, when there is less runoff, the nutrients have a better chance of being taken up by the plants or immobilized by the soil.

A study of three corn-cropped watersheds in southwestern Iowa found nutrient losses to be greatest between April and June, with 54 percent of the annual loss of soluble nitrogen, 90 percent of the sediment-associated nitrogen, and 85 percent of the sediment phosphorus being discharged from the watershed during that period.[50] This is the period when fertilizer application is heaviest, when there are many storms, and when erosion is heaviest.[51]

Nitrogen. This basic nutrient is found in agricultural soils in several different physical and chemical forms—particulate or dissolved, strongly adsorbed or free, organic or inorganic, cationic or anionic. All are involved in a complex and dynamic cycle of individual nitrogen molecules continually changing from one form to another.[52]

Plants require a higher volume of nitrogen than of any other plant nutrient found in agricultural soils. The major sources of naturally occurring nitrogen in soils are plant residues, legumes (whose roots contain bacteria that convert nitrogen from the air into organic compounds), manure, and other biological residues. The nitrogen from these sources initially enters the soil in organic, insoluble, complex forms such as proteins and nucleic acids.* At any one time, most of this nitrogen remains in an organic, insoluble form and must be carried off the land in suspension or attached to sediment.[53]

However, as the nitrogen-containing compounds decompose, bacteria break down (mineralize) the nitrogen to produce inorganic ionic compounds—first, ammonium salts (through ammonification); then, sometimes, nitrites and nitrates (through nitrification). Ammonification occurs rapidly in moist, warm soils that are rich in nutrients and organic matter and support flourishing microbial populations, both plants and animals. The mineralization process usually ends with ammonification; each year, only 2 or 3 percent of organic nitrogen becomes a nitrite or nitrate.[54] Instead, ammonium and intermediate products of ammonification often are assimilated by the microbial population and reconverted to organic compounds. In addition, ammonium may be taken up by crop plants before it is nitrified.

Any organic form of nitrogen can be carried from soil through surface runoff and can be mineralized after it enters a waterway. In addition, soluble forms of inorganic nitrogen can be carried away by runoff. When nitrogen is carried off the soil in the ammonium stage, it usually is bound to soil particles, although it can also be dissolved in the water. In soils that contain clays that expand when they get wet, ammonium may be trapped within the clay's structure so that it is relatively unavailable either to promote crop growth or to cause problems if the soil is carried off the farm. (This form, called fixed ammonium, can account for 3 to 8 percent of the total soil nitrogen.[55]) Nitrates and nitrites are highly soluble and mobile.

The most common form of commercial fertilizer, used about 45 percent of the time, is anhydrous ammonia. Nitrogen solutions (combinations of ammonium, nitrates, and urea) are responsible for about 25 percent. Urea applied separately accounts for another 14 percent.[56] Anhydrous ammonia and nitrogen solutions are inorganic, water soluble, and applied directly to the soil in a liquid form so that they are readily available for uptake by plants. Plain urea, by contrast, is organic and is usually applied in granular form on the soil surface.

Not surprisingly, the proportion of inorganic nitrogen in soil is likely to increase significantly after commercial fertilizer is applied. This may explain the seemingly contradictory results produced by experiments

*Manure, however, also can contain significant amounts of ammonia and other inorganic forms of nitrogen.

measuring the organic/inorganic nitrogen ratio in the streams of eroding watersheds. In one study, organic nitrogen accounted for 71 percent of the total nitrogen found in eight streams, ammonium for 24 percent, and nitrates and nitrites for only 5 percent.[57] By contrast, a nationwide study of 928 watersheds found that nitrates and nitrites comprised almost 80 percent of total nitrogen concentrations in streams predominantly draining agricultural land, compared to only 18 percent in streams draining forested watersheds.[58]*

Phosphorus. Like nitrogen, phosphorus is present in soils in many different forms: some is organic, some inorganic; some is soluble, some insoluble. The effects of phosphorus depend on the particular physical and chemical form it is in.

About two-thirds of the phosphorus occurring in natural soil is inorganic,[59] but the exact balance of inorganic to organic phosphorus shifts constantly with the size of the microbial population and the amount of residue and other organic components of the soil.[60] Decomposition and mineralization change organic phosphorus (such as is found in decaying vegetation) into dissolved, inorganic phosphate ions that are biologically available for plant uptake. To complete the cycle, bacteria incorporate those phosphate ions and synthesize them into organic phosphorus, a process known as immobilization.[61]

Both organic and inorganic phosphorus may be either soluble or insoluble.[62] Soluble phosphorus is easily carried away in runoff; insoluble phosphorus is mainly transported adsorbed† onto sediment particles. All of the soluble phosphorus is available for uptake by plants, but only 20 to 40 percent of the particulate phosphorus is similarly available.[63] In one study, for instance, only 20.7 percent of the sediment phosphorus in agricultural drainage water was in a biologically

---

*The differences could also relate to the amount of erosion occurring on the fields. With high erosion rates, the total amount of nitrogen carried off would be high. However, nitrites and nitrates should compose a relatively small proportion of this, because they account for a relatively small share of the total nitrogen in the soil. When erosion is low, the amount of nitrogen (including nitrates and nitrites) should be much lower, but the proportion made up by the inorganic forms higher since, being soluble, they are carred off more easily. The difference in percentages should not obscure the fact that the total amount of nitrates and nitrites coming off rapidly eroding land will most likely be greater than the amount coming off noneroding land. It is just that large amounts of organic nitrogen are also carried off these lands—and much of this can undergo the mineralization process after it enters the waterways.

†Adsorption involves a physiochemical process by which a thin layer of a substance, often only a few molecules thick, adheres to the surface of soil particles.

available form.[64]

Substances can desorb* as well, with the rate of net adsorption or desorption depending upon the concentration of the substance in the surrounding water. The more phosphorus there is in the water, the more adsorption will occur; conversely, the less concentrated the phosphorus in the water solution, the more likely it is that the adsorbed phosphorus will be re-released into the solution.[65] Fine clay particles, with a high ratio of surface area to volume, adsorb soil nutrients most readily, although the degree of bonding depends on the mineral composition (and ion exchange capacity) of the clay. Phosphate ions bind more strongly to clays containing magnesium and calcium than to those with sodium.[66]

Since some of the phosphorus occurring in natural soils is not in a form readily available to plants, fertilizers are applied to augment the natural levels of available phosphorus. Most of the phosphorus in commercial fertilizers is in the form of phosphoric oxide, a water-soluble, inorganic form. Once the fertilizer comes into contact with the soil, it reacts quickly with elements in the soil to form less soluble phosphate salts and may be strongly adsorbed by the soil particles. The acidity of the soil influences the particular reaction that occurs: in acid soils, the phosphate compounds are rapidly converted to iron and aluminum phosphates; in alkaline soils, they form calcium phosphates.[67] The rate at which phosphates come into contact with the active soil components is determined by factors such as the method of fertilizer application (surface application or subsurface injection), soil moisture, temperature, and the amount of tillage and ground cover.[68] (Phosphorus found in manure acts differently from most chemical fertilizers because it contains an especially high concentration of readily soluble organic phosphorus.[69])

Other Nutrients. Many elements besides nitrogen and phosphorus can be classified as nutrients essential to plant growth. Some of these—potassium and certain trace metals such as copper, zinc, iron, and magnesium—are applied to soil in chemical fertilizers. However, there appears to be relatively little information about what physical and chemical changes influence the biological availability of these nutrients to plants and bacteria after they are applied to a field.

Transport of Nutrients

Each year, water and wind together carry an estimated 4.2 million tons

*Desorption is the reverse of absorption or adsorption.

of nitrogen and 1.0 million tons of phosphorus off U.S. agricultural lands.[70] According to one estimate, as much as 50 kilograms of nutrients per metric ton of sediment are potentially washed off farmers' fields.[71]

The rate at which a nutrient is transported depends both on its method of transport and the environmental factors around it. If the nutrient is sediment-bound, its movement from a field is influenced by the same factors that determine the amount of sediment eroded: rainfall intensity, sediment particle size, field slope, vegetative cover, and tillage practices. For a soluble nutrient, the rate of transport depends primarily on the amount and timing of rainfall and runoff. Vegetative cover also can increase the amount of a soluble nutrient in runoff because the cover may allow nutrients to be washed off a field before they come into contact with the soil.

Scientists disagree about the ability of nutrients to cause damage during transport.[72] It is generally thought, however, that sediment-bound phosphorus is less damaging to water quality than soluble phosphorus is, because the insoluble form is less biologically available for uptake by algae. Still, there may be so much more sediment phosphorus than dissolved phosphorus that the former may cause more total damage. Phosphorus from nonpoint sources in rural watersheds has been estimated to be 80 percent particulate and 20 percent water-soluble.[73]

The sediment-transport process is complicated by the ability of nutrients to desorb from, as well as adsorb onto, sediment particles. For instance, the amount of available phosphorus in surface waters depends partly on the source of the sediment. If the sediment is from fertile topsoil, the associated phosphorus can desorb relatively easily from the soil particles, increasing the concentration of dissolved phosphorus. However, if the sediment is composed of infertile subsoils or other phosphorus-deficient materials, it can adsorb dissolved phosphorus already in the waterway.[74]

Nutrient Enrichment. The most easily eroded component of soil—particles of clay—also tends to be the portion of soil that binds most actively with nutrients. Consequently, eroded sediment often contains a much higher concentration of nutrients than does the surface soil from which it comes. A study of sediments discharged from Indiana's Black Creek into the Maumee River showed that the transported sediment contained about three times the total phosphorus concentration and about five times the total nitrogen concentration as the soil left behind.[75]

The difference in nutrient concentration between eroded soil and source soil is expressed in a "nutrient-enrichment ratio." Such a ratio usually correlates well with the amount of sediment enrichment that occurs, with both decreasing as erosion increases.* A study of an Iowa agricultural watershed showed that approximately 85 percent of its total phosphorus yield was associated with sediment transport.[77] Similarly, data from the Huron River in Ohio showed a direct relationship between the amount of sediment and the amount of phosphorus in the river.[78] And, in New York State, high flow conditions accounted for approximately 75 percent of the annual phosphorus loss from an agricultural watershed, even though such conditions occurred only 10 percent of the time.[79]

Several other studies have documented the significance of agricultural land as a source of nutrients in waterways. In the Great Lakes region, land-use activities (for instance, farming, pasture, forestry) were found to contribute from one-third to one-half of the total phosphorus load in the Great Lakes, especially Erie and Ontario; cropland was identified as the major source of nonpoint phosphorus loads.[80] A 1977 U.S. Environmental Protection Agency study of watersheds in the United States reported that combined mean concentrations of total phosphorus and nitrogen were nearly nine times greater in streams draining agricultural land than in streams draining forested areas.[81]†

This problem of nutrient runoff from agricultural lands can be expected to increase. Although fertilizer consumption may not grow at the high rates of past years, it is likely to continue growing nevertheless. Limits to the amount of fertilizer that farmers can economically apply to crops do exist, but in many cases they have not been reached. In some situations, more fertilizer will be needed to offset the continuing loss of natural supplies of nutrients in good soils, to coax high yields from marginal soils, and to compensate for the effects that excessive erosion and destructive farming practices can have on productivity.

Nutrients in Lakes and Reservoirs. Many nutrients appear not to settle in rivers during transport but, instead, eventually to reach a larger body of water. A study of the Maumee and Cuyahoga river basins during spring storms in 1975, for instance, found that most, if not all, of the sediment-associated phosphate in the rivers was flushed into Lake Erie.[82] Once these contaminants reach a lake or ocean, however, a variety of complex chemical and biological interactions determine

---

*Of course, even though higher erosion causes a lower nutrient-enrichment ratio, it also results in greater *overall* nutrient transport.[76]

†The relative importance of agriculture can vary substantially from one watershed to another. In addition, municipal and industrial discharges, feedlots, and natural vegetation can be significant sources of nutrient runoff.

their ability to alter water quality and affect aquatic life.

One such factor is the speed with which sedimentation occurs.* Of particular importance is the length of time sediment particles remain suspended in the "photic" zone—that is, the top layer of water, which has sufficient light for plant photosynthesis. Fine clay particles usually require at least four hours to settle out of a five-meter photic zone, and currents, winds, and waves often keep them suspended for a day or more. During this period, the nutrients bound to this sediment can promote algal growth.[84]

Other factors influencing the impact of nutrients on aquatic environments include the amount of sediment delivered to a water body, the temperature of the water, the amount of light, the species and populations of algae present, the availability of other essential nutrients, and the acidity or alkalinity of the water. Water acidity, in turn, can be influenced by materials carried into the water, the composition of soils in the watershed, and the materials forming the stream or lake bed, as well as by outside influences such as acid precipitation.[85]

Even after deposition occurs, nutrients can be recirculated in a water body. Phosphorus, for instance, can be released back into water by the process of desorption. According to one study, 18 to 65 percent of the total inorganic phosphorus contained in a wide range of bottom sediments is exchangeable.[86] Because the rate of desorption increases as the concentration of dissolved phosphorus decreases, experiments have shown that as algae take up phosphorus and immobilize it, enough replacement phosphate is released from the sediment to maintain a concentration sufficient for continued algal growth.[87]†

---

*One study monitoring the inflow and outflow from a Missouri reservoir in an agricultural region found that the concentration of dissolved orthophosphate (the dominant form of soluble phosphorus) in the outflow averaged only 48 percent of that in the inflow. The rest of the orthophosphate may have been taken up by algae in the reservoir or adsorbed by sediment deposited in or carried out of the reservoir. In general, water leaving a reservoir contains less particulate phosphorus, exchangeable phosphorus, and orthophosphorus than the water entering the reservoir.[83] However, the reduction in phosphorus may be less than the reduction in total sediment because the phosphorus tends to bind to the more mobile, clayey particles, which are least likely to settle out in the reservoir.

†This process is more complicated than a simple chemical equilibrium process. The majority of exchangeable inorganic sediment phosphorus comes from the "nonapatite fraction," which is the phosphorus mainly associated with iron and, less so, aluminum sediment components. Under anaerobic conditions, the release of inorganic phosphorus is accelerated because, in the absence of oxygen, iron hydroxides are reduced to other compounds, thereby releasing the phosphate. Thus,

Sediment nutrients may also be returned to circulation by the action of aquatic organisms. Some marine plants, for instance, take up nutrients from the seabed and release them from their leaves into surrounding waters. Similar plant species have been found in freshwater lakes.[89] For instance, in Shagawa Lake in Minnesota, the amount of phosphorus recycled by plants equals the amount removed from sewage effluent at a nearby municipal waste-treatment facility.[90] In addition, some bottom-feeding species of fish have been found to recycle sediment nutrients by excreting phosphate and ammonia in significant amounts. An average population of carp in a Minnesota lake, for example, can release enough nutrients to produce massive algal growth.[91]

The precise amount of lake-sediment nutrients available for re-release is hotly debated. Successive depositions of sediment effectively bury earlier deposits, possibly making all but the top few inches of sediment unavailable for algal growth. However, the churning caused by bottom-dwelling fish, burrowing animals, dredging, and other activities can bring buried nutrients back to the bed's surface.

Pesticides

Pesticides* are the second major category of contaminants associated with soil erosion. Like nutrients, they are transported from cropland either by being adsorbed onto eroding soil particles or dissolved in runoff water. Pesticides also can leach into groundwater or evaporate into the atmosphere.

For sediment-bound pesticides, biological availability is determined by the type of pesticide and sediment involved. Weakly adsorbing pesticides, such as the phenoxy acid herbicides (for instance, 2,4-D), readily desorb into water and become more biologically available. However, other pesticides, such as paraquat, become unavailable because they strongly adsorb onto the internal surfaces of certain types of clay; then, once the clay particles are eroded and suspended in water, the moisture causes them to expand, trapping the pesticides inside. Highly insoluble pesticides, such as the dinitroaniline and urea her-

when the algae originally fed by the nutrients sink to the bottom and die, their decomposition depletes the oxygen in the bottom sediments, which in turn increases phosphate release and plant productivity in a continuing cycle of eutrophication (explained in chapter 4).[88]

*Pesticide is the general term for all chemicals used to kill pests. Insecticides (used to kill insects) are one type of pesticide. Herbicides (used to kill plants), fungicides (used to kill fungi), and rodenticides (used to kill rodents) are others.

bicides, that are adsorbed on the external surfaces of sediments may be gradually released for uptake by aquatic plants and animals.[92]

Some pesticides can accumulate in plant or animal tissues, so that they are more concentrated in living tissues than in the surrounding environment. For instance, pesticides with low water solubilities are lipid (fat) soluble: not readily excreted once they are ingested, they accumulate in the fatty tissues of fish, animals, and humans. The organochlorines (including DDT [dichloro-diphenyl-trichloro-ethane], aldrin/dieldrin, heptachlor, chlordane, and toxaphene) are the most dramatic examples. Such substances can also be "biomagnified" through the food chain. As lower organisms are eaten by their predators, the pesticides stored by the lower organisms are passed through to the consuming animal, increasing in concentration at each level of the food chain. And, it has been shown, the less water-soluble a pesticide is, the more it tends to accumulate in plants and animals.[93]

Pesticides in Agriculture

Hundreds of different active pesticide ingredients are currently used in the United States, and 15 to 30 new ones are introduced each year. Between World War II and 1975, the number of pesticides increased from 100 to over 900.[94] Most are synthetic organic compounds. The most common types of insecticides currently in use are organophosphates (methyl and ethyl parathion, malathion, diazinon, acephate, and phorate) and carbamates (carboryl and carbofuran). Organochlorines used to be very common, but all have been banned for most agricultural uses in the United States. Nevertheless they still persist in the environment, though in steadily diminishing amounts.[95] More than 180 basic types of herbicides exist, although until recently only a few such as atrazine and alachlor were widely used.[96]*

Pesticide application on agricultural land has increased dramatically over the last 30 years, primarily because of the increased use of herbicides (figure 2.5). In 1956, only 11 percent of corn acreage and 5 percent of cotton acreage were treated with herbicides; by 1982, the percentages had increased to 95 percent and 97 percent, respectively.[98] In 1977, over 95 percent of the corn and soybeans grown in Iowa received herbicides; over 55 percent of the corn also received insecticides.[99] Between 1976 and 1982, herbicide use nationally increased 18 percent

---

*Atrazine and alachlor accounted for one-half of all herbicides used in farming in 1976, followed by 2,4-D, trifluralin, and butylate.[97] Fyphosphate (Roundup) is rapidly becoming one of the more important herbicides.

on corn, 56 percent on soybeans.[100]

Most pesticides that are now in surface waters are herbicides, insecticides, and fungicides that originally were intended for fields and orchards.* The pesticides often never even reach their intended target areas. For instance, two-thirds of the insecticides used in agriculture are applied by aircraft, a means of application that can result in 50 to 75 percent of the chemical drifting downwind and settling in drainage ditches and streams, on nontarget crops, and on noncultivated land.[101]

Figure 2.5
U.S. Production of Pesticides, by Type, 1960–1982

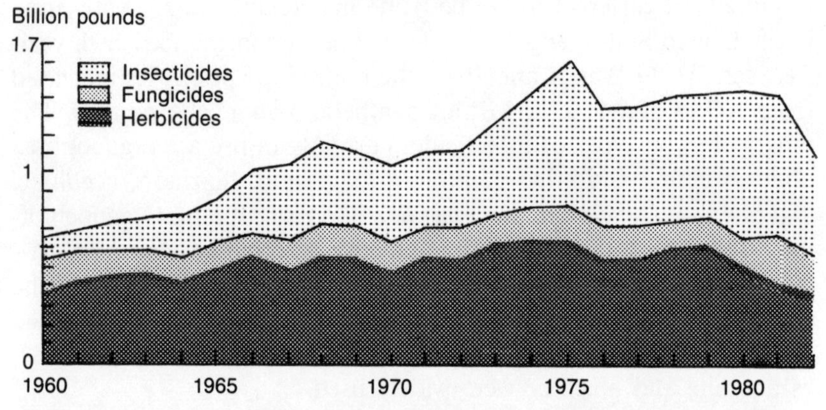

Source: U.S. International Trade Commission, *Synthetic Organic Chemicals 1970* (Washington, D.C.: U.S. Government Printing Office, 1979), and previous annual issues; and U.S. International Trade Commission, *Synthetic Organic Chemicals: United States Production and Sales, 1982* (Washington, D.C.: U.S. Government Printing Office, 1983).

*Not all pesticide pollution is attributable to cropland application, however. Other sources include urban runoff, forestland and pastureland runoff, accidental spills, careless disposal practices, hazardous-waste dumps, and discharges at production sites.

Transport of Pesticides

An estimated 360 tons of pesticides are carried away from U.S. agricultural lands by wind and water each year.[102] Several factors determine the amount of a pesticide that is transported into and through a waterway. One such influence is the period of time between the pesticide's application and the next significant rainfall. One study found that, for most pesticides, total annual losses from agricultural fields were 0.5 percent or less of the amounts applied, unless severe rainfall occurred within one or two weeks of application.[103] Another study found that the quantity of the herbicides atrazine and simazine transported from fields increased with the amount of runoff and was inversely related to the length of time between application and the rainfall that caused the runoff. Even under heavy rainfall conditions, no pesticide loss in the experiment exceeded 6 percent of the amount applied.[104] Under such unusual conditions as a heavy storm on the day of application, however, runoff losses may exceed 15 percent.[105]

The form in which a pesticide is applied to a field—solid or liquid—also can affect the amount carried off. Pesticides applied to the surface in powder form are easily dislodged from the soil surface by rain, while liquid pesticides are more likely to adhere to the plants or the soil.

In addition, transport rates vary according to whether a pesticide is applied to the surface of the soil or directly incorporated into the upper soil layer. Surface-applied pesticides are more easily lost through runoff, simply because pesticides incorporated into the soil are less available to be washed off. For instance, an experiment with fonofos found that incorporation resulted in significantly lower concentrations of the insecticide in runoff than did surface application.[106]

The chemical and physical characteristics of the pesticides themselves also affect how, and the extent to which, they are transported from the application site. Chlorinated hydrocarbons and other insoluble pesticides usually are transported on soil particles rather than dissolved in water.[107] More soluble pesticides are much more likely to be transported by being dissolved in runoff. Until agricultural usage of organochlorine pesticides was banned, eroded soil probably was the most important method of pesticide transport. But, with the increased use of more soluble organic phosphate insecticides as well as the rapid growth in the use of more soluble herbicides, runoff—not erosion—is increasing as a predominant cause of pesticide contamination of surface waters.[108]

Moreover, a pesticide's mobility is influenced by its propensity to be adsorbed by soil. One study found that losses of fonofos, which is highly adsorptive, averaged only 20 percent of those for alachlor.[109] Over 50 percent of the fonofos losses that did occur were with sediment. Pesticides with strong organic cations (positive charges), such as diquat and paraquat, are relatively immobile because of their strong attraction for clay minerals, which are high in negatively charged ions. Weakly adsorbed pesticides, such as the anionic (negatively charged) herbicides dicamba, picloram, amiben, or 2,4-D, are highly mobile.[110] Nonionic pesticides (those lacking a strong positive or negative charge), such as the organochlorine and organophosphorous insecticides, are less likely to be adsorbed by inorganic particles than are ionic pesticides.[111]

Pesticides eventually decompose, and usually (but not always) they become less toxic in the process. However, much remains to be learned about how decomposition, and the end products resulting from it, can be influenced by the chemical environment in which it occurs. Some significant factors in certain circumstances could be the presence or absence of sunlight, oxygen, other reactive chemical compounds, and microbes.

The longer a pesticide persists before decomposing, the more of it is available for transport during successive rainfalls. Whether or not pesticides present off-farm problems often depends on how soon they are carried off the land after they are applied. Many of the newer pesticides, such as the carbamate and organophosphate insecticide groups, can break down within weeks or sometimes days into relatively nontoxic components, even though they are acutely toxic when applied.[112] Other pesticides—among them, organochlorine insecticides—persist in soils and aquatic environments for several years: DDT requires 11 years for a 95 percent breakdown into harmless components; dieldrin, 9.7 years; and chlordane, 4.2 years.[113] The importance of the rate of chemical breakdown also has been demonstrated by a study that compared the persistence and runoff losses of atrazine and alachlor and that found that 150 percent more of the more persistent atrazine was carried off.[114]

A pesticide's relative persistence determines not only how much of it is carried off the field but also how long it can cause problems once it is transported. If the newer, shorter-lived pesticides decompose before they are transported far from their application site, any environmental damage probably will be fairly localized. The impacts of phosvel and carbofuran, for instance, are apparently limited to the upstream areas of a watershed because of their low persistence.[115] In contrast, highly persistent organochlorines such as DDT and aldrin/dieldrin are still found in sediments, often far from the places where the pesticides were originally used, many years after their use was halted.[116] In addition, some pesticides can last much longer in water than in soil. For instance, aldicarb, an insecticide with a half-life in soil of a few weeks, apparently can persist up to 20 years in water.[117]

Another major role in determining pesticide losses is played by soil characteristics. Highly permeable, sandy soils reduce runoff but, at the same time, their permeability allows pesticides to seep into the groundwater or to be carried by drainage tiles into streams. Aldicarb has frequently been found in groundwater, particularly in locations like Long Island, where sandy soils allow rapid infiltration.[118] On the other hand, soils that are highly compacted (for instance, by farm machinery) decrease pesticide infiltration, producing significantly higher pesticide concentrations in surface runoff water and sediment.[119]

The chemical composition of the soil also influences the extent to which a pesticide is adsorbed and immobilized by soil particles. The adsorption potential of soils high in organic matter may be 10 to 1,000 times greater than that of inorganic soils. As a result, soils high in organic content will yield less pesticide to runoff (unless serious erosion occurs). Similarly, in a waterway, the concentration of dissolved pesticides can be decreased if the suspended or deposited sediments have a high organic content and are, therefore, able to adsorb the pesticides.[120] In addition, a soil's acidity or alkalinity can influence the tendency of some pesticides to be adsorbed.[121]

Clear distinctions cannot always be made, however, between pesticides transported dissolved in runoff and those transported by sediment. Because adsorption is reversible, some pesticides transported from a field primarily by sediment will, upon reaching surface waters, desorb and be carried in solution.

Moreover, the damages that pesticides can cause once they have been carried off a field do not depend only on the total amounts transported. A pesticide's biological availability is another important factor. However, knowledge of these many factors affecting pesticide transport and biological impact is far from complete.

Other Contaminants

Eroded soil carries with it an array of contaminants other than nutrients and pesticides. Soil's organic materials are subject to the same basic mechanisms of erosion and transport as its mineral components, with a few notable exceptions; being lighter in weight, and less tightly bound to the rest of the soil, the organic particles can be selectively eroded from a field. Crop residues and manure, if left on the soil surface instead of being plowed under, can be washed off a field even before erosion begins. They carry along whatever contaminants are tightly bound to them and can continue to adsorb dissolved pesticides and nutrients after reaching a waterway.

The oxygen-consuming microbial decomposition of organic materials continues after the substances reach a waterway, consuming the dissolved oxygen in the water. Thus, minimum dissolved oxygen levels may occur well downstream—often in a lake or reservoir fed by the stream. Such oxygen depletion can suffocate fish and other aquatic organisms. In addition, the decomposition process can release adsorbed contaminants back into solution in the water.*

Even if they settle to the bottom and become buried in the sediments of a stream, lake, or reservoir, organic particles, like sediment-bound nutrients, are still able to cause problems. When waves or currents resuspend bottom sediments, decomposition can begin again, increasing the consumption of dissolved oxygen manyfold. For instance, when the sediment in one pool along the Illinois River was disturbed, the amount of oxygen consumed in the pool because of decomposition processes in the sediment increased more than 600 percent over demand levels for calm water.[122]

Bacteria and Viruses

Many beneficial bacteria occur naturally in soil. Some convert atmospheric nitrogen into available plant nutrients; others play an essential role in decomposing organic matter and breaking down chemical compounds that could be toxic in their original form. Bacteria are responsible for the eventual decomposition of most materials into such basically innocuous components as carbon dioxide, methane, water, and inorganic minerals, as well as nitrogen and phosphorus.

Pathogenic bacteria and viruses, however, also can be present. Animal droppings, manure, sewage spread on fields, and tissues of diseased plants all contaminate soil with pathogens. Soil erosion can then carry these bacteria to streams and rivers to contaminate surface waters.[123] One study estimated that up to 6 percent of the fecal bacteria produced by grazing livestock is washed from pastures and into waterways.[124] Such problems are likely to be particularly acute near animal feedlots.

*On the other hand, it may also contribute to the decomposition of toxic contaminants into less harmful products.

Bacteria are fairly mobile, moving with the water contained within soil. They also can be adsorbed onto finely textured soil particles and be transported from fields by the selective erosion of those particles.[125] Some evidence indicates that attachment to soil particles will enhance bacterial survival after they reach a waterway.[126]

The survival time of bacteria in soil can vary from a few hours to years. Several factors in addition to adsorption influence this survival time: kind of bacterium, soil type, pH, temperature, sunlight, rainfall, and predation of other soil organisms (figure 2.6).[127] For instance, intestinal bacteria, which can create major health problems (such as hepatitis) if they reach surface waters, may persist in soil for two to three months.[128] They are more likely to survive in moist, clayey, or peaty soils and will survive longer at lower temperatures.

Viruses can survive in soil and fresh water for several days or for several months. The same factors determine viruses' longevity as influence bacteria's. Viruses readily adsorb to soil particles, allowing them to be transported in association with sediment. This process does not reduce their infectiousness. Sunlight kills viruses, so their viability during transport is linked to the amount of sunlight that penetrates the water (which is, in turn, affected by the turbidity of the water).[129]

Figure 2.6
Survival of Bacteria and Viruses in Soil

Factor		Comment
pH	Bacteria	Shorter survival in acid solids (pH 3 to 5) than in neutral and calcareous soils
	Viruses	Insufficient data
Predation by soil microflora	Bacteria	Increased survival in sterile soils
	Viruses	Insufficient data
Moisture content	Bacteria and viruses	Longer survival in moist soils and during periods of higher rainfall
Temperature	Bacteria and viruses	Longer survival at low temperatures
Sunlight	Bacteria and viruses	Shorter survival at the soil surface
Organic matter	Bacteria and viruses	Longer survival or regrowth of some bacteria when sufficient amounts of organics are present

Source: Vladimir Novotny and Gordon Chesters, *Handbook of Nonpoint Pollution Sources and Management* (New York: Van Nostrand Reinhold, 1981), table 6–6, p. 258.

Metals

Trace metals—such as copper, manganese, iron, zinc, boron, and cadmium—are necessary to crop growth and therefore are sometimes applied in chemical fertilizers as micronutrients (as was mentioned earlier). Often, they occur naturally in soils. Some are released into the ground, as dead plants and animals decompose. However, although they are not widespread contaminants of agricultural soils, trace metals can be present in the soil in amounts that threaten the downstream environment. Like the sediment with which they are associated, these metals are far more likely to be eroded from cultivated cropland than from natural, uncultivated areas. One Florida study, for instance, found that the highest concentrations of dissolved metals in a canal were located adjacent to cultivated areas.[130]

A potentially more serious source of contamination of waterways with eroded metals is the disposal on agricultural land of sludge produced by municipal wastewater-treatment plants. This sludge, which is a source of valuable organic matter and nutrients, can also contain toxic metals and other contaminants, particularly if it comes from facilities treating industrial wastes. And, since sludge has a tendency to fill the holes in coarsely textured soil, it consequently increases runoff and the transport of pollutants to surface waters.[131] Although the application of sewage sludge to agricultural lands in the United States is relatively uncommon, two-thirds of the sludge still is disposed of on land somewhere.[132]

In addition, small amounts of metals may be included in pesticides and fertilizers applied to agricultural soils. Some pesticides use a metal as a toxic agent or other component. In such cases, the breakdown of the pesticide after application may release the metal and increase the environmental hazard. Fertilizer compounds for certain crops also may contain metals. Cauliflower, for instance, malforms if it lacks sufficient boron. These metal additives can be carried off land either directly with the fertilizer or indirectly with the plant residues.

Salts

"Dissolved solids"—salt ions such as sulfate, chloride, bicarbonate, sodium, magnesium, and calcium that dissolve easily in water—are another major pollutant category associated with soil erosion and runoff from agricultural lands.

These salts are particularly common in more arid areas. They can reach surface waters by an assortment of paths. Intense rainstorms and runoff pick up salts in areas of severe erosion, transporting them in dissolved runoff or as part of the sediment, to subsequently dissolve in a receiving waterway. Water dissolves salts as it moves downward through the soil, carrying them into the groundwater, which may subsequently seep back into a surface waterway. In arid regions, soil moisture tends to migrate upward and evaporate, leaving a crust of salt on the soil surface to then be carried off by runoff and erosion.[133] These salts can also be carried away by irrigation return flows.

Most salt ions remain dissolved in the water as they are transported down a stream or river. They can, however, react with other chemicals and precipitate or become adsorbed onto suspended sediment particles. Dissolved solids can seriously affect aquatic life in the receiving waters and hamper the growth of crops irrigated downstream. In addition, some salts—in particular, magnesium and calcium—exacerbate water hardness, causing scaling and corrosion and reducing the lathering and effectiveness of soaps.

Up to 40 percent of the salinity in western streams and lakes may be contributed by human activity,[134] including erosion and runoff from the tillage of naturally saline soils. The damage that the dissolved salts cause depends on their concentration. Therefore, although more of them reach surface waters during storms, the greater volume of water that is also in a waterway during and after heavy rainfall can actually decrease the salt concentration. However, in drier regions, problems can arise farther downstream, if evaporation causes the waterway's salinity to increase.

LINKAGE TO OFF-FARM IMPACTS

The background discussions presented in this chapter provide the bases for identifying the types of off-farm impacts of soil erosion that one might expect to find. The erosion and runoff processes create a set of potential pollutants—the resulting sediment and associated substances. These pollutants are transported into and through waterways, where they can directly or indirectly cause certain effects. These effects, in turn, directly or indirectly create various costs (or benefits) for society.

This chapter has concentrated on the first of these links—the relationship between the erosion process and the generation, transport, and fate of potential pollutants. The next three chapters focus on the second and third links—from pollutant to effect, and from effect to cost (or benefit).

REFERENCES

1. R. P. Beasley, *Erosion and Sediment Pollution Control* (Ames, Iowa: Iowa State University Press, 1972), p. 10.

2. Ibid.

3. Harry R. Leach, "Soil Erosion," in Oscar E. Meinzer, ed., *Hydrology* (New York: McGraw-Hill Book Co., 1942), pp. 609, 667.

4. Hugh Hammond Bennett, *Soil Conservation* (New York: McGraw-Hill Book Co., 1939), p. 67.

5. U.S. Department of Agriculture, *Predicting Rainfall Erosion Losses,* Agricultural Handbook no. 537 (Washington, D.C.: U.S. Government Printing Office, 1978), p. 4.

6. U.S. Department of Agriculture, Agricultural Research Service, *Proceedings of the Workshop on Estimating Erosion and Sediment Yields on Rangelands—Tucson, Arizona, March 7-9, 1981,* Western Series no. 26 (Oakland, Calif.: U.S. Department of Agriculture, 1982), pp. 166-86. This has an extensive discussion of cover-management factors for rangeland and forestland.

7. U.S. Department of Agriculture, *Predicting Rainfall Erosion Losses,* p. 5.

8. Ibid., p. 5.

9. Ibid., pp. 9-11.

10. Ibid., pp. 12-13.

11. Ibid., p. 15.

12. Ibid., p. 18, 19.

13. Ibid., p. 19.

14. Ibid., p. 19, figures 6 and 7.

15. S. C. Kimes, J. L. Baker, and H. P. Johnson, "Sediment Transport from Field to Stream: Particle Size and Yield" (American Society of Agricultural Engineers, St. Joseph, Mich., 1979), pp. 19-21.

16. David B. Baker, "Fluvial Transport and Processing of Sediment and Nutrients in Large Agricultural River Basins," prepared for U.S. Army Corps of Engineers, Lake Erie Wastewater Management Study, Buffalo, N.Y., and U.S. Environmental Protection Agency, Office of Research and Development, Environmental Research Laboratory (1982), pp. 114-15.

17. Vladimir Novotny and Gordon Chesters, *Handbook of Nonpoint Pollution: Sources and Management* (New York: Van Nostrand Reinhold Co., 1981), p. 188.

18. M. Gordon Wolman, "Changing Needs and Opportunities in the Sediment Field," *Water Resources Research* 13 (1977):51.

19. Thomas E. Davenport and Julie J. Oehme, "Soil Erosion and Sediment Delivery in the Blue Creek Watershed, Pike County, Illinois," Division of Water Pollution Control (Illinois Environmental Protection Agency, Springfield, Ill., 1982), p. 17.

20. Stanley W. Trimble, "Changes in Sediment Storage in the Coon Creek Basin, Driftless Area, Wisconsin, 1853 to 1975," *Science* 214 (October 9, 1981).

21. Leonard P. Gianessi and Henry M. Peskin, "Analysis of National Water Pollution Control Policies: 2. Agricultural Sediment Control," *Water Resources Research* 17, no. 4 (1981):805.

22. Novotny and Chesters, *Handbook of Nonpoint Pollution,* pp. 196-97.

23. F. E. Dendy, "Sediment Yield from a Mississippi Delta Cotton Field," *Journal of Environmental Quality* 10, no. 4 (1981):485-86.

24. Ibid., p. 485.

25. U.S. Department of Agriculture, Soil Conservation Service, *SCS National Engineering Handbook,* sect. 3, *Sedimentation* (Washington, D.C.: U.S. Department of Agriculture, 1973), pp. 2-7.

26. Oswald Rendon-Herrero, "Estimation of Washload Produced on Certain Small Watersheds," *Journal of the Hydraulics Division, American Society of Civil Engineers* 100 (1974):843.

27. Luna B. Leopold, M. Gordon Wolman, and John P. Miller, *Fluvial Processes in Geomorphology* (San Francisco: W. H. Freeman and Co., 1964) pp. 189-91, 192-93.

28. Vito A. Vanoni, ed., *Sedimentation Engineering,* prepared by the American Society of Civil Engineers Task Committee for the preparation of the Manual on Sedimentation of the Sedimentation Committee of the Hydraulics Division (New York: American Society of Civil Engineers, 1975), pp. 228-29.

29. Vanoni, *Sedimentation Engineering,* p. 177.

30. James Guilluly, Aaron C. Waters, and A. O. Woodford, *Principles of Geology,* 3d ed. (San Francisco: W. H. Freeman and Co., 1968), p. 75; and Vanoni, *Sedimentation Engineering,* p. 178.

31. M. A. Churchill, "The Silt Investigations Program of the Tennessee Valley Authority" (Proceedings of the First Federal Interagency Sedimentation Conference, May 6-8, 1947), p. 53.

32. Leopold, Wolman, and Miller, *Fluvial Processes in Geomorphology,* p. 183; and Guilluly, Waters, and Woodford, *Principles of Geology,* p. 210.

33. Vanoni, *Sedimentation Engineering,* pp. 179.

34. Ibid., pp. 183-89.

35. Guilluly, Waters, and Woodford, *Principles of Geology,* p. 210; Baker, "Fluvial Transport and Processing of Sediment and Nutrients," p. 4; Dendy, "Sediment Yield from a Mississippi Delta Cotton Field"; and Rendon-Herrero, "Estimation of Washload Produced on Certain Small Watersheds."

36. Rendon-Herrero, "Estimation of Washload Produced on Certain Small Watersheds," p. 837; and Thomas H. Cahill et al., *Phosphorus Dynamics and Transport in the Brandywine Basin,* Technical Paper no. 3, (Tri County, Pa.: Tri-County Conservancy of the Brandywine, 1975), p. 21.

37. John H. Greiner, Jr., *Erosion and Sedimentation by Water in Texas—Average Annual Rates Estimated in 1979* (Austin, Tex.: Texas Department of Water Resources, 1982), p. 73.

38. Robert Coats et al., *Landsliding, Channel Change, and Sediment Transport in Zayante Creek and the Lower San Lorenzo River, 1982 Water Year and Implications for Management of the Stream Resource* (Berkeley, Calif.: Center for Natural Resource Studies, 1982), p. 67.

39. Bern Keating, *The Mighty Mississippi* (Washington, D.C.: National Geographic Society, 1971), p. 145.

40. U.S. Army Corps of Engineers, Mississippi River Commission, New Orleans District, "Atchafalaya Basin Floodway System—Louisiana Feasibility Study," vol. 1, "Main Report and Final Environmental Impact Statement—January 1982" (U.S. Army Corps of Engineers, Vicksburg, Miss., 1982), p. 9.

41. Ven Te Chow, ed., *Handbook of Applied Hydrology: A Compendium of Water-Resources Technology* (New York: McGraw-Hill Book Co., 1964), pp. 17-28.

42. F. E. Dendy, "Sedimentation in the Nation's Reservoirs," *Journal of Soil and Water Conservation* 23, no. 4 (1968):135.

43. Leo R. Beard, *Sediment Effects of Headwater Reservoirs: Trinity River, Texas,*

Technical Report no. CRWR-163, prepared by the University of Texas at Austin, College of Engineering, Center for Research in Water Resources, for the U.S. Department of Agriculture, Soil Conservation Service (Austin, Tex.: University of Texas at Austin, 1979), p. 5.

44. U.S. Department of the Interior, Geological Survey, "Effects of Dams on Rivers Are Documented" (Washington, D.C.: U.S. Geological Survey), p. 60.

45. Michael N. Beaulac and Kenneth H. Reckhow, "An Examination of Land Use: Nutrient Export Relationships," *Water Resources Bulletin* 18, no. 6 (1982):1020.

46. D. R. Timmons, R. F. Holt, and J.H. Catterell, "Leaching of Crop Residue as a Source of Nutrient in Surface Runoff," *Water Resources Research* 6 (1975):1367; and Raymond C. Loehr, "Characteristics and Comparative Magnitude of Non-point Sources of Pollution," *Journal of the Water Pollution Control Federation* 46, no. 8 (1974):1852.

47. U.S. Department of Agriculture, "Commercial Fertilizers: Consumption for the Year Ended June 30, 1983" (U.S. Department of Agriculture, Washington, D.C., 1983), p. 4.

48. Loehr, "Characteristics and Comparative Magnitude of Non-point Sources of Pollution," p. 1852.

49. Bert Boln and Robert B. Cook, eds., *The Major Biogeochemical Cycles and Their Interactions* (New York: John Wiley and Sons, 1983), pp. 284, 292.

50. E. E. Alberts, G. E. Schuman, and R. E. Burwell, "Seasonal Runoff Losses of Nitrogen and Phosphorus from Missouri Valley Loess Watersheds," *Journal of Environmental Quality* 7, no. 2 (1977):207.

51. Ibid., p. 203.

52. D. R. Keeney, "Transformations and Transport of Nitrogen," in Frank W. Schaller and George W. Bailey, eds., *Agricultural Management and Water Quality* (Ames, Iowa: Iowa State University Press, 1983), contains a thoroughly succinct summary of what is known about the nitrogen cycle in agricultural soils.

53. T. Logan, "Forms and Sediment Associations of Nutrients (C, N, and P), Pesticides and Metals," in H. Shear and A. E. P. Watson, eds., *Proceedings of a Workshop on the Fluvial Transport of Sediment-Associated Nutrients and Contaminants* (Windsor, Ontario: International Joint Commission, 1977), p. 182.

54. Mingteh Chang, Jack D. McCullough, and Frank A. Roth "Nitrogen Yields and Land Use in Southern Streams," (Proceedings of the Symposium on Watershed Management '80, American Society of Civil Engineers, Boise, Idaho, 1980), p. 958.

55. Logan, "Forms and Sediment Associations of Nutrients, Pesticides and Metals," p. 182.

56. U.S. Department of Agriculture, Crop Reporting Board, "Commercial Fertilizers," p. 21.

57. Chang, McCullough, and Roth, "Nitrogen Yields and Land Use in Southern Streams," p. 957.

58. James M. Omernik, *Nonpoint Source—Stream Nutrient Level Relationships: A Nationwide Study,* prepared for U.S. Environmental Protection Agency, Office of Research and Development, Corvallis Environmental Research Laboratory, EPA-600/ 3-77-105 (Corvallis, Oreg.: U.S. Environmental Protection Agency, 1977), p. 1.

59. Darrell W. Nelson and Terry J. Logan, "Chemical Processes and Transport of Phosphorus," in Schaller and Bailey, *Agricultural Management and Water Quality,* p. 68.

60. Ibid., pp. 68-69.

61. Novotny and Chesters, *Handbook of Nonpoint Pollution,* pp. 218-19.

62. Nelson and Logan, "Chemical Processes and Transport of Phosphorus," pp. 66-67.

63. Timothy J. Monteith, Melanie P. Baise, and Rose Ann C. Sullivan, "Environmental and Economic Implications of Conservation Tillage Practices in the Great Lakes Basin" (Great Lakes Environmental Planning Study, Contribution no. 20, 1981), p. 29.

64. R. A. Dorich and D. W. Nelson, "Algal Availability of Soluble and Sediment Phosphorus in Drainage Water of the Black Creek Watershed," *Journal of the Water Pollution Control Federation* 9, no. 4 (1980):557.

65. Nelson and Logan, "Chemical Processes and Transport of Phosphorus," pp. 74-75.

66. Novotny and Chesters, *Handbook of Nonpoint Pollution,* p. 219-20.

67. Ibid., p. 221.

68. E. M. White, "Possible Clay Concentration Effects on Soluble Phosphate Contents of Runoff," *Environment Science Technology* 15 (1981):104.

69. G. Y. Reddy, E. O. McLean, and G. D. Hoyt, "Effects of Soil, Cover Crop, and Nutrient Source on Amounts and Forms of Phosphorus Movement under Simulated Rainfall Conditions," *Journal of Environmental Quality* 7, no. 1 (1978):54.

70. Gianessi and Peskin, "Analysis of National Water Pollution Control Policies," p. 804.

71. M. L. Quinn, ed., "Strategies for Reducing Pollutants from Irrigated Lands in the Great Plains" (Nebraska Water Resources Center, University of Nebraska-Lincoln, Lincoln, Nebr., and U.S. Environmental Protection Agency, Robert S. Kerr Environmental Research Laboratory, Ada, Okla., 1982), p. 43.

72. Nelson and Logan, "Chemical Processes and Transport of Phosphorus," pp. 85-86.

73. Monteith, Baise, and Sullivan, "Environmental and Economic Implications," p. 29.

74. G. Stanford, C.B. England and A. W. Taylor, *Fertilizer Use and Water Quality,* prepared for U.S. Department of Agriculture, Agricultural Research Service, 41-168 (Washington, D.C.: U.S. Department of Agriculture, 1970), p. 18.

75. E. J. Monke et al., "Sediment and Nutrient Movement from the Black Creek Watershed," *Transactions of the American Society of Agricultural Engineering* 24 (1981):395.

76. A. N. Sharpley, "The Enrichment of Soil Phosphorus in Runoff Sediments," *Journal of Environmental Quality* 9 (1980):521; Susan B. Klein, *State Soil Erosion and Sediment Control Laws: A Review of State Programs and Their Natural Resource Data Requirements* (Denver, Colo.: National Conference of State Legislatures, 1980); and Michael R. Overcash and James M. Davidson, eds., *Environmental Impact of Nonpoint Source Pollution* (Ann Arbor, Mich.: Ann Arbor Science Publishers, 1980), pp. 283-87.

77. T. C. Daniel and R. C. Wendt, "Nonpoint Pollution: Sediment and Phosphorus in Agricultural Runoff" (University of Wisconsin, Cooperative Extension Program, 1979), p. 6.

78. T. H. Cahill, R. W. Pierson, Jr., and B. Cohen, "The Evaluation of Best Management Practices for the Reduction of Diffuse Pollutants in an Agricultural Watershed," in Raymond C. Loehr et al., eds., *Best Management Practices for*

Agriculture and Silvaculture (Proceedings of the 1978 Cornell Agricultural Waste Management Conference) (Ann Arbor, Mich.: Ann Arbor Science Publishers, 1979), p. 468.

79. Daniel and Wendt, "Nonpoint Pollution: Sediment and Phosphorus in Agricultural Runoff," p. 6.

80. International Joint Commission, *Pollution in the Great Lakes Basin from Land Use Activities—Summary* (Windsor, Ontario: International Joint Commission, 1980), p. 2.

81. Omernik, *Nonpoint Source-Stream Nutrient Level Relationships, p. 1.*

82. Logan, "Forms and Sediment Associations of Nutrients," p. 177.

83. J. D. Schreiber and D. L. Rausch, "Suspended Sediment-Phosphorus Relationships for the Inflow and Outflow of a Flood Detention Reservoir," *Journal of Environmental Quality* 8, no. 4 (1979):511, 512, 514.

84. P. J. Huettl, R. C. Wendt, and R. B. Corey, "Prediction of Algal-Available Phosphorus in Run-off Suspensions," *Journal of Environmental Quality* 8, no. 1 (1979):131.

85. Novotny and Chesters, *Handbook of Nonpoint Pollution,* p. 127.

86. D. E. Armstrong, "Phosphorus Transport Across the Sediment-Water Interface," in *Lake Restoration* (Proceedings of a national conference, August 22-24, 1978, Minneapolis, Minn.), EPA 440-5-79-001 (Washington, D.C.: U.S. Government Printing Office, 1979), p. 169.

87. Ibid., p. 170.

88. Ibid.

89. Joseph Shapiro, "The Need for More Biology in Lake Restoration," in *Lake Restoration,* p. 161.

90. G. B. Lie, "Phosphorus Cycling by Freshwater Macrophytes: The Case of Shagawn Lake" (Contribution no. 184 from the Linmnology Research Center, University of Minnesota); cited in Ibid., p. 162.

91. Shapiro, "The Need for More Biology in Lake Restoration," p. 162.

92. J. B. Weber, "Geochemistry of Sediment/Water Interactions of Nutrients, Pesticides and Metals, including Observations on Availability," in Shear and Watson, *Proceedings of a Workshop on the Fluvial Transport of Sediment-Associated Nutrients and Contaminants,* p. 251.

93. R. L. Metcalf and J. R. Sanborn, "Pesticides and Environmental Quality in Illinois," *Illinois Natural History Survey Bulletin* 31, no. 9 (1975):401.

94. Ibid., p. 382.

95. The Conservation Foundation, *State of the Environment: An Assessment at Mid-Decade* (Washington, D.C.: The Conservation Foundation, 1984), pp. 50-54.

96. Bette Hileman, "Herbicides in Agriculture", *Environmental Science and Technology* 16, no. 12 (1982):646A.

97. Ibid..

98. Council on Environmental Quality, *Environmental Trends* (Washington, D.C.: U.S. Government Printing Office, 1981), p. 94; Economic Research Service, *Inputs,* October 1983, p. 4.

99. Overcash and Davidson, *Environmental Impact of Nonpoint Source Pollution,* p. 300.

100. Theodore R. Eichers, *Farm Pesticide Economic Evaluation, 1982,* prepared for U.S. Department of Agriculture, Economics and Statistics Service, AERS no. 464 (Washington, D.C.: U.S. Government Printing Office, 1981), p. 9.

101. Council on Environmental Quality, *Environmental Trends,* p. 92.

102. R. D. Wauchope, "The Pesticide Content of Surface Water Draining from Agricultural Lands—A Review," *Journal of Environmental Quality* 7, no. 4 (1978):459; and Council on Environmental Quality, *Environmental Trends,* p. 92.

103. R. D. Wauchope and R. A. Leonard, "Maximum Pesticide Concentrations in Agricultural Runoff: A Semiempirical Prediction Formula", *Journal of Environmental Quality* 9, no. 4 (1980):1362.

104. D. R. Dudley and J. R. Karr, "Pesticides and PCB Residues in the Black Creek Watershed, Allen County, Indiana, 1977-1978," *Pesticides Monitoring Journal* 13 (1980):156.

105. Overcash and Davidson, *Environmental Impact of Nonpoint Source Pollution,* p. 276.

106. Overcash and Davidson, *Environmental Impact of Nonpoint Source Pollution,* p. 293.

107. Novotny and Chesters, *Handbook of Nonpoint Pollution,* p. 233.

108. National Research Council, Committee on Impacts of Emerging Agricultural Trends on Fish and Wildlife Habitat, *Impacts of Emerging Agricultural Trends on Fish and Wildlife Habitat* (Washington, D.C.: National Academy Press, 1982), pp. 272-73.

109. Overcash and Davidson, *Environmental Impact of Nonpoint Source Pollution,* p. 276.

110. Ibid.

111. Novotny and Chesters, *Handbook of Nonpoint Pollution,* p. 232.

112. North Carolina Soil and Water Conservation Commission, Division of Land Resources, and the 208 Agricultural Task Force, *Water Quality and Agriculture: A Management Plan* (unpublished, n.d.), p. 14.

113. Metcalf and Sanborn, "Pesticides and Environmental Quality in Illinois," p. 395.

114. Overcash and Davidson, *Environmental Impact of Nonpoint Source Pollution,* p. 276.

115. Logan, "Forms and Sediment Associations of Nutrients, Pesticides and Metals," p. 213.

116. International Joint Commission, *Pollution in the Great Lakes Basin from Land Use Activities,* p. 39.

117. "Water Contaminated by Pesticide Spurs Growing Concern in Florida," *New York Times,* March 28, 1983.

118. Ward Sinclair, "America's Pesticide Use Raises New Safety Fears," *The Washington Post,* January 30, 1983.

119. Overcash and Davidson, *Environmental Impact of Nonpoint Source Pollution,* p. 276.

120. Logan, "Forms and Sediment Associations of Nutrients, Pesticides and Metals," p. 213.

121. Overcash and Davidson, *Environmental Impact of Nonpoint Source Pollution,* p. 232.

122. Illinois Institute for Environmental Quality, "Task Force on Agriculture Nonpoint Sources of Pollution—Final Report," December 1978, p. 38.

123. Novotny and Chesters, *Handbook of Nonpoint Pollution,* p. 257.

124. Chesapeake Research Consortium, "Non-Point Source Studies on Chesapeake Bay: III. Relationship between Bacterial Contamination and Land Use

in the Rhode River Watershed, and Survival Studies of 'Streptococcus faecalis' in the Estuary,'' prepared for National Science Foundation, CRC no. 56 (Springfield, Va.: National Technical Information Service, 1977), p. 71.

125. Novotny and Chesters, *Handbook of Nonpoint Pollution,* p. 259.

126. Chesapeake Research Consortium, "Non-Point Source Studies on Chesapeake Bay: III,'' p. 73.

127. Novotny and Chesters, *Handbook of Nonpoint Pollution,* p. 257.

128. Ibid., p. 259.

129. Ibid., p. 258.

130. R. W. Skaggs et al., *Effect of Agricultural Land Development on Drainage Waters in the North Carolina Tidewater Region,* (Springfield, Va.: National Technical Information Service, 1980), p. 86.

131. G. K. Rutherford, "Anthropogenic Influences of Sediment Quality at a Source,'' in Shear and Watson, *Proceedings of a Workshop on the Fluvial Transport of Sediment-Associated Nutrients and Contaminants,* p. 98.

132. The Conservation Foundation, *State of the Environment: An Assessment at Mid-Decade* (Washington, D.C.: The Conservation Foundation, 1984), p. 132.

133. Pueblo Regional Planning Commission, "Executive Summary for Pueblo County: Agricultural Water Quality Assessment of Lower Fountain Creek,'' COL-208 (Pueblo Area Council of Governments, Pueblo, Colo., 1982), p. 7.

134. J. P. Law, Jr., and H. Bernard, "Impacts of Agricultural Pollutants in Water Uses,'' *Transactions of the American Society of Angricultural Engineering* 1354 (1975):474-78; cited in U.S. Environmental Protection Agency, *Planning Guide for Evaluating Agricultural Nonpoint Source Water Quality Controls,* EPA-600/ 3-82-021 (Athens, Ga.: U.S. Environmental Protection Agency, 1982), p. 7.

3. Impacts of Sediment

The inert soil particles picked up in the erosion process can cause a wide range of impacts between the time they leave a field and the time they reach their permanent resting place in an ocean, in a lake, or somewhere else along the way. Many of these impacts occur while the particles are in a waterway such as a stream, river, lake, or reservoir and are called in-stream impacts. Others occur before the particles reach a waterway or after they leave it, either in floodwater or in water withdrawn for municipal use, irrigation, or some other beneficial purpose. These are called off-stream impacts.

IN-STREAM IMPACTS

Soil particles can cause in-stream impacts either while they are suspended in water or after they have settled out of it. Either way, the sediment can affect a wide range of uses of the waterway. When suspended, sediment can affect the viability of various forms of aquatic life, reduce the value of the water for such uses as recreation, and increase maintenance costs at hydroelectric and other facilities that use water without removing it from a waterway. Settled sediment can cause serious problems for aquatic life by covering food sources, hiding places, and nesting sites. It can also clog navigation channels, reduce the capacities of stream channels and reservoirs, and, as indicated in chapter 2, extensively change a waterway's structure and ecology. In some cases, an impact on one water use may affect other uses. For instance, a decrease in the viability of the aquatic ecosystem will affect a water body's recreational uses.

Biological Impacts

Sediment affects aquatic organisms at every link in the food chain, from microscopic algae to valuable game fish. It causes damage either directly, through physically or biologically affecting an organism itself, or indirectly, through destroying the organism's required habitat. In

Kansas and some other areas, the sediment is thought to be "the major single pollutional factor influencing aquatic systems."[1] In England, polluted streams average 2 to 5 fish per 1,000 feet, while unpolluted streams average 16 to 27 fish per 1,000 feet.[2]

Evaluating the relative importance of sediment's effects on aquatic plants and animals is difficult, however. The presence and distribution of a particular organism is influenced by a wide variety of other factors: the responses of the specific species, the ecological composition of the aquatic community, the composition of the sediment, and variations in the physical and chemical environment (such as differences in water temperature, current velocity, pH, dissolved oxygen and carbon dioxide concentrations, nutrients, metals, and pesticides). Moreover, many of these environmental factors are themselves affected by the presence of sediment.[3] On one hand, sediment containing a high proportion of organic matter tends to exert high oxygen demand on an aquatic environment, consuming dissolved oxygen that would otherwise be available to living organisms. On the other hand, the same organic material may adsorb harmful pesticide residues in the water, making them less available for fish to consume.

Suspended sediment most often negatively affects a waterway by increasing its turbidity—that is, by reducing its transparency and the amount of sunlight that can penetrate it. The particles that settle may travel down a stream as bed load. This bed load, even though it usually accounts for a small proportion of a waterway's total sediment load,[4] is an important cause of sedimentation-related problems. Of course, as explained in chapter 2, settled particles of soil can be repeatedly suspended and deposited. As a result, the particles may contribute intermittently to both turbidity and sedimentation problems.

Water-quality criteria for turbidity usually are based either on the effects of turbid water on aquatic life[5] or on water requirements for public drinking-water supplies,[6] on the assumption that avoiding harm in either of these uses will adequately protect other water uses as well. However, because of problems that exist in measuring turbidity,* the

---

*Turbidity can be measured approximately in different ways, but no current system can accomodate itself to the numerous varied effects that can be caused by differences in material, lighting, and sediment-associated problems. One of the systems used to estimate turbidity is the "Jackson Turbidity Unit." This system relates the way in which light is absorbed or scattered by a sample of water to the concentration of fine silica that produces an equivalent effect.[7] Alternatively, turbidity can be measured as the maximum depth of water through which an observer can see a secchi disk, a black-and-white plate 20 centimeters in diameter,[8] or by the relative weight of the materials suspended in the water.

U.S. Environmental Protection Agency (EPA), for instance, does not propose any specific quantitative standard in its current recommendations regarding the problem.[9]

No water-quality criteria related specifically to sedimentation have been developed, but EPA and others have considered the problems sedimentation causes when they have recommended turbidity guidelines to protect fish and other aquatic life.[10] EPA's current recommendations assume that keeping turbidity low enough to protect against excessive reduction of sunlight penetration will also protect against sedimentation problems.

Turbidity

Many different materials—among them, silt, algae, and bacteria—can increase a water body's turbidity, and each will affect the water's transparency differently. Water transparency also can depend on the color and angle of the light entering the water. Even in completely clear lake water, only about 40 percent of the light penetrates to a depth of one meter.[11]*

Turbidity keeps sunlight from penetrating to the deeper layers of a water body by both reflecting and absorbing the light. In slow-moving or still water, this increases the temperature stratification and interferes with the normal mixing of oxygen-rich surface waters with those deeper down.[13] As algae and other organic materials sink downward, their decomposition uses up what dissolved oxygen there is in the lower layers. Because the oxygen is not being replenished through mixing from the surface, less is available to support fish and other oxygen-dependent organisms. Water bodies with high fish populations are much more likely to be associated with low levels of turbidity than are bodies with medium or nonexistent fish populations.

Decreased light penetration also lowers primary production—the photosynthetic production by green plants of organic material from carbon dioxide and water. In aquatic environments, primary production is limited to the euphotic zone, the water depth at which sufficient light exists for photosynthesis. Increased turbidity, by reducing the euphotic zone, greatly decreases the quantity of phytoplankton (microscopic plants such as algae) produced. One researcher found that, among Oklahoma farm ponds, clear ponds (less than 25 parts per million [ppm] of suspended sediment) produced 8 times as much

*Other problems caused by suspended sediment, such as abrasion and clogging of fish gills, are insufficiently reflected in turbidity measurement methods.[12]

plankton as did ponds with intermediate suspended-sediment levels (25-100 ppm) and 12.8 times as much plankton as did ponds with high suspended-sediment levels (over 100 ppm).[14] Similarly, a reduction in the amount of light reaching the bottom of a body of water hinders the growth of benthic, or bottom-dwelling, macrophytes (rooted aquatic plants).

A reduction in aquatic plants may have effects that reverberate through an aquatic ecosystem. Zooplankton and fish larvae depend on phytoplankton for their food. Weed fauna—small snails and insects—graze on macrophytes themselves or on the attached scum of bacteria and algae.[15] Weed fauna, in turn, are preyed upon by fish and by dragonfly and damselfly nymphs. All these organisms, aquatic and nonaquatic, are affected by a reduction in phytoplankton and macrophytes.

Because turbid water reduces visibility within a water body, it can affect the viability of some aquatic populations. For example, different forms of weed fauna and other aquatic invertebrates can be affected in various ways. By inhibiting algal production, turbidity can eliminate cladocerans and copepods, minute crustaceans that get their food by filtering microorganisms from the water. It also can starve caddis flies, which spin nets in water to capture food, by clogging the nets with sediment of little or no nutritional value.[16] Some species of planorbid snails will not lay eggs in turbid water; others lay eggs that do not develop.[17]

Increased turbidity also is a major problem for fish that rely on sight, rather than tactile senses, to find prey, since it greatly shortens a predator's reactive distance (the greatest distance at which it can locate its prey). A 50 percent cut in reactive distance results in a 90 percent reduction in the total volume of water that a fish can survey for prey at any given moment.[18] (A corresponding effect of turbid water is to allow young fish and other prey to escape their predators.)

Reduced visibility also is a problem for fish that depend on visual cues for courtship and spawning behavior.[19] If sunfish, sticklebacks, killifish, and some darters cannot see each other, they may be unable to carry out the complex courtship behavior that seems to be required for their successful reproduction. Other fish with simpler spawning behavior—species in which the male and female look alike or who tend to reproduce at night, relying on chemical and tactile clues to find each other—will be less affected by this problem.

Turbidity apparently can affect reproduction in other ways, too. Although no evidence has been found indicating that fish will not pass through turbid waters on their way to spawning grounds, turbidity does affect reproductive timing. The spawning of largemouth bass, for instance, has been delayed in muddy ponds for as much as 30 days after spawning took place in nearby clear ponds. And, in some cases, female bass have been observed to resorb their eggs in turbid water rather than to spawn.[20]

Even if eggs are successfully laid, turbidity and its effects on sunlight and oxygen availability may interfere with normal hatching, reduce larva size at hatching, cause developmental abnormalities, and make it more difficult for the many fish larvae that rely on visual detection of plankton during their initial feeding stages to find enough food.

Direct, lethal effects of sediment on adult fish are seldom observed in nature, although they have been demonstrated in the laboratory. For an adult fish to be killed directly by suspended sediment, the concentration of sediment must be high enough to clog the organism's gill passages, depriving the fish of oxygen and causing death by suffocation.[21] Although tolerance levels vary among species, the high sediment levels found to be lethal in lab experiments are unusual in nature, particularly for any extended period of time.

However, sublethal effects of sediment on fish are widely observed. Adult fish exposed to high levels of suspended solids have suffered gill damage and abrasion.[22] Some evidence also indicates that sediment-induced stress reduces growth rates and decreases resistance to diseases. Rainbow trout, for instance, develop a higher incidence of fin rot when exposed to turbid water for several months.[23] And growth and development of reproductive capability in largemouth bass is stunted when suspended solids reduce their available food.[24] A study correlating fish yield with turbidity found that clear farm ponds in Oklahoma produced 1.7 to 5.5 times the total weight of fish obtained from more turbid ponds.[25] This result is probably due to the reduced plankton growth in the more turbid ponds.

Still, assessing how much direct damage suspended sediment actually causes in natural settings is difficult: most fish avoid turbid conditions if at all possible and, if they are healthy, wash away sediment particles by mucus production. Moreover, fish populations often are reduced by sediment's indirect impacts—decreases in food supply, habitat, and reproductive success—or by other contaminants in the water or food chain before sediment's direct effects can be detected.[26]

Sedimentation

Sedimentation may be delivering an even more serious blow than turbidity to aquatic plant and animal life.

Plants can be affected by sedimentation in a variety of ways. Large sediment loads dumped in lakes and reservoirs by a storm can obliterate macrophytes. For instance, increased silt loads are considered responsible for the loss of lotus beds in the Saint Clair River and Lake Erie's Sandusky Bay.[27] Many plants that require firm soil in which to root cannot take hold in soft, unstable sediment covering a river or lake bottom. Abrasion and scouring by bed-load sediment can destroy photosynthetic organisms. Less drastically, light, fine-particled sediment can settle onto and cover the leaves of aquatic plants, thereby reducing their photosynthetic ability.[28]

These reductions in aquatic plant populations have serious repercussions throughout an aquatic environment. The importance of phytoplankton and macrophytes to that environment has been explained earlier. Such vegetation also is important to terrestrial animals. In marshes and lake margins, for instance, it is a food source for many sources of waterfowl.

Aquatic invertebrates, particularly benthic ones, are directly affected by sedimentation.* Deposition of sediment in streams fills the spaces between rocks, destroying the organisms' habitats. Stone flies, mayflies, and caddis flies, if they have not been smothered, will leave an affected area, drifting downstream in search of more favorable habitat.[30] However, other benthos (bottom-dwelling organisms), such as mollusks and tube worms, are less mobile and cannot escape degraded water conditions.[31] For them, elimination of habitat often spells their elimination as well. Excessive amounts of fine sediment can also reduce the food sources of benthic invertebrates by eliminating the interrock spaces in which leaf material and detritus are deposited.[32]

Research on fingernail clams—a widely distributed, important food source for ducks, channel catfish, carp, and bullheads—provides examples of these impacts. These clams respond to sediment-laden water by closing their shells, thereby keeping any food from being absorbed in their filters. In one experiment, sediment loads reduced the clams' feeding time by 50 percent.[33] This may explain why, in reaches of the Illinois River suffering from sedimentation, clam populations have been significantly reduced or have disappeared entirely and why the carp there are measurably thinner and smaller than normal.[34] In clearer reaches, the clams provide half of the carp's diet.

---

*Such organisms are also affected by hydrological and geomorphological factors such as less stable channels and higher flood flows associated with increased erosion (see chapter 2). The high water flows and heavy sediment loads combine to scour river and stream bottoms, dislodging benthic organisms and sweeping them downstream.[29]

Similarly, siltation destroys the habitat of oysters and hard clams, which need to attach themselves to firm bottoms free of heavy mud.[35] One study estimated that sedimentation has destroyed one-half of the fish and oyster spawning grounds in the upper estuary of the Chesapeake Bay.[36] Sedimentation is also thought to have destroyed large portions of the mussel populations on the Mississippi, Tennessee, and Ohio rivers earlier this century.[37] The mussels' disappearance correlated roughly chronologically with the intensification of row-crop farming in those regions.

But sedimentation not only affects plants and benthic animals. By killing submerged vegetation and by filling in pools, "fishing holes," and deep channels, sedimentation eliminates features upon which many aquatic species depend for protective cover, especially to protect newly hatched fry. Gars, pike, suckers, sunfish, and various minnows and darters attach their eggs to plants. Suckers, sunfish, and minnows also require spaces among rocks along a waterway's bed, as do lampreys, perch, and catfish. Females of some species have been observed to resorb their eggs when heavy sediment killed vegetation or eliminated rocky bottoms.[38] The few fish whose eggs float on the water's surface are not affected by high sediment loads.[39] However, most fish deposit their eggs on the bottom. For instance, in the Great Lakes, all fish species except the freshwater drum are in the latter category.[40]

Fish species characterized by complex territorial and nest guarding behavior are also more affected by sedimentation than species that engage in minimal parental care of their eggs and young. Territorial fish have trouble recognizing and defending their territories when sediment obscures landmarks on the river bottom. Tippecanoe darters desert their territorial holdings when stream silt loads increase after storms.[41] Some egg-tending species continually fan their nests to provide the eggs with a constant flow of oxygen-rich water; when heavy sediment loads drive the parents from their nests, the eggs suffocate.

The incubation stage of eggs is a particularly susceptible period in the life cycle of fish. The small size of many warmwater-fish eggs makes smothering by settling sediment a real possibility in shallow windswept reservoirs or lakes and in unstable streams where bank sloughing and heavy sediment loads are common. Walleye, northern pike, and yellow perch eggs all have demonstrated significant mortality from sediment burial.[42]

Sediment particles also can damage fish eggs by abrasion and physical damage to the outer membrane, the chorion. Damaged chorions allow fungus spores to become established, a common problem on eggs of warmwater fishes in sediment-laden streams.[43]

Sediment's destructive effects on the eggs and fry of salmon and trout are particularly well documented.[44] Salmon lay their eggs in redds (nests) dug down into gravelly stream bottoms and cover the eggs with the gravel during incubation. The spaces between gravel particles must be large enough to hold the eggs and to allow water carrying dissolved oxygen to circulate freely. Once the eggs hatch, the fry must be able to wriggle upward through the gravel and out into the stream. Sediment deposition after the eggs have been laid can be disastrous. Not only do many eggs suffocate, but, even if they hatch, the fry are trapped within the streambed. In one study of chinook salmon, for instance, the siltation of as little as 15 to 30 percent of the intergravel spaces resulted in 85 percent mortality of the eggs and fry.[45] And another study concluded that removing the natural silt in the gravel beds of salmon spawning areas in Alaska would increase their productivity about fivefold.[46]

Such high mortality rates severely reduce the size of future adult fish populations. Even under normal conditions, very few fish survive to adulthood. Usually fewer than 10 percent of the eggs laid by brook trout, for instance, produce fish that are still alive by the end of their first summer of life.[47] Heavy sedimentation makes these odds much worse.

The net effects of sedimentation are well documented in studies of actual fish populations in streams and ponds. In New Brunswick, researchers concluded that sedimentation and chemical contamination were the major reasons that fish and benthic organisms were less abundant in streams near farms than in streams flowing through natural and clear-cut forests.[48] Many species of Great Lakes fish have experienced significant population reductions partly attributable to increased sedimentation.[49]

Formerly clear, gravel-bottomed streams in the western United States also have seen dramatic declines in their sport fish populations. Salmon and steelhead trout populations in California's San Lorenzo River plummeted from an average annual run of 20,000 steelhead and 2,500 to 10,000 silver salmon in 1964 to 3,000 and 182, respectively, in 1978.[50] Sediment-caused habitat damage—sedimentation of spawning gravels, filling of pools, water-quality degradation, and stream-flow diversion—is the principal culprit.

Heavy sedimentation also can substantially affect the species composition of the populations that do survive. The warmer water temperatures of shallow, silted lakes are intolerable to certain kinds of fish (figure 3.1). Salmon, trout, smallmouth bass, walleye, sauger, and some darter species will be replaced by more tolerant (or "rough") fish such as carp, goldfish, gizzard shad, and carpsuckers.[51] In North Carolina, the State Wildlife Resources Commission holds sediment responsible for the domination of such rough fish species in many of that state's stream segments.[52]

Fish Adaptation to Turbidity and Sedimentation

Although the effects of sediment on fish reproduction and development are potentially very great, some species have adopted traits that allow them to avoid the problem (figure 3.2). In many areas in the West, the heavy sediment loads resulting from long-term geological erosion have led to the dominance of species whose reproductive activities are carried on outside times of highest turbidity. Such coolwater species as northern pike and perch spawn during a brief interval in early spring, thereby avoiding the high turbidity found in late spring. Pioneer species in northern headwaters (including stonerollers, creek chub, and orangethroat darters) also complete their spawning before late spring rains bring high, muddy waters.[53]

However, such traits are not common among native fish species in the East. As indicated in chapter 1, the heavy sediment loads in eastern streams have occurred relatively recently, with the clearing of forests and intensive farming following European settlement. Because eastern streams had traditionally small sediment loads, the majority of native warmwater species in the East tended to spawn in the late spring and early summer rainy season, when high river flows stimulated migration, provided suitable spawning habitat by flooding and clearing channel bottoms, and provided enough fresh water-flow for egg development. The high sediment loads that now accompany these spring flows have severely affected the reproductive success of those species.

One result of this situation is that Illinois waterways are now dominated by numerous sediment-tolerant fish whose origins are in the turbid plains streams of the West.[54] For instance, some of the dominant species are short-lived minnows and darters that lay several egg clutches per season, thereby extending the reproductive period beyond the rainy season and reducing the probability of an entire year class being eliminated in a flood.

Recreation

Sediment can greatly reduce the suitability of water for recreational uses. Some of this impact is direct. For instance, silt-laden water is more dangerous for activities such as swimming and boating.[55] Other problems arise indirectly from such biological impacts of turbidity and

Figure 3.1
Warmwater Fishes that are Intolerant of Suspended Solids and Sediment

Species		Effect		Impact From	
Common name	Scientific name	Spawning	General	Suspended Solids	Sediment
Lampreys					
Chestnut lamprey	*Ichthyomyzon castaneus*	X			X
Sturgeons					
Lake sturgeon	*Acipenser fulvescens*	X	X		X
Paddlefishes					
Paddlefish	*Polyodon spathula*	X	X		X
Gars					
Shortnose gar	*Lepisosteus platostomus*		X		X
Bowfins					
Bowfin	*Amia calva*	X		X	
Mooneyes					
Mooneye	*Hiodon tergisus*		X	X	
Pikes					
Northern pike	*Esox lucius*	X		X	X
Muskellunge	*Esox masquinongy*		X	X	
Carps and minnows					
Redside dace	*Clinostomus elongatus*		X		X
Tonguetied minnow	*Exoglossum laurae*		X		X
Cutlips minnow	*Exoglossum maxillingua*		X		X
Bigeye chub	*Hybopsis amblops*		X	X	X
Streamline chub	*Hybopsis dissimilis*		X		X
Gravel chub	*Hybopsis x-punctata*		X		X
Horneyhead chub	*Nocomis biguttatus*	X			X
River chub	*Nocomis micropogon*		X	X	X
Pallid shiner	*Notropis amnis*		X	X	
Bigeye shiner	*Notropis boops*		X	X	X
Common shiner	*Notropis cornutus*		X		X
Pugnose minnow	*Notropis emiliae*		X	X	X
Blackchin shiner	*Notropis heterodon*		X	X	X
Blacknose shiner	*Notropis heterolepis*		X	X	
Spottail shiner	*Notropis hudsonius*		X	X	X
Ozark minnow	*Notropis nubilus*		X		X
Rosyface shiner	*Notropis rubellus*		X	X	X
Sand shiner	*Notropis stramineus*		X	X	X
Weed shiner	*Notropis texanus*		X		X
Topeka shiner	*Notropis topeka*		X		X
Mimic shiner	*Notropis volucellus*		X	X	
Suckers					
Highfin carpsucker	*Carpiodes velifer*		X	X	
Blue sucker	*Cycleptus elongatus*		X		X
Creek chubsucker	*Erimyzon oblongus*		X	X	
Lake chubsucker	*Erimyzon sucetta*		X	X	X
Northern hog sucker	*Hypentelium nigricans*		X	X	X
Harelip sucker	*Lagochila lacera*		X	X	X

Species		Effect		Impact From	
Common name	Scientific name	Spawning	General	Suspended Solids	Sediment
Spotted sucker	*Minytrema melanops*		X	X	
River redhorse	*Moxostoma carinatum*		X	X	X
Black redhorse	*Moxostoma duquesnei*		X	X	X
Greater redhorse	*Moxostoma valenciennesi*		X	X	X
Bullhead catfishes					
Blue catfish	*Ictalurus furcatus*		X	X	X
Stonecat	*Noturus flavus*		X		X
Carolina madtom	*Noturus furiosus*		X	X	
Tadpole madtom	*Noturus gyrinus*		X	X	X
Brindled madtom	*Noturus miurus*		X	X	
Scioto madtom	*Noturus trautmani*		X	X	X
Flathead catfish	*Pylodictis olivaris*		X	X	X
Trout-perches					
Trout-perch	*Percopsis omiscomaycus*		X		X
Killifishes					
Blackstrip topminnow	*Fundulus notatus*		X	X	
Silversides					
Brook silverside	*Labidesthes sicculus*		X	X	
Sticklebacks					
Brook stickleback	*Culaea inconstans*		X	X	
Sunfishes					
Rock bass	*Ambloplites rupestris*		X	X	
Pumpkinseed	*Lepomis gibbosus*		X	X	X
Longear sunfish	*Lepomis megalotis*		X	X	
Smallmouth bass	*Micropterus dolomieui*	X	X	X	X
Largemouth bass	*Micropterus salmoides*		X	X	X
Perches					
Crystal darter	*Ammocrypta asprella*		X		X
Western sand darter	*Ammocrypta clara*		X		X
Eastern sand darter	*Ammocrypta pellucida*		X		X
Greenside darter	*Etheostoma blennioides*		X		X
Iowa darter	*Etheostoma exile*		X	X	
Tippecanoe darter	*Etheostoma tippecanoe*		X		X
Banded darter	*Etheostoma zonale*		X		X
Yellow perch	*Perca flavescens*	X	X	X	X
Logperch	*Percina caprodes*		X	X	X
Channel darter	*Percina copelandi*		X	X	X
Gilt darter	*Percina evides*		X	X	X
Blackside darter	*Percina maculata*		X	X	
Slenderhead darter	*Percina phoxocephala*		X	X	X

Source: Robert J. Muncy et al., *Effects of Suspended Solids and Sediment on Reproduction and Early Life of Warmwater Fishes: A Review*, prepared for U.S. Environmental Protection Agency, EPA 600/3-79-042 (Washington, D.C.: U.S. Government Printing Office, 1979), table 7, p. 69–71.

Figure 3.2
Warmwater Fishes that are Tolerant of Suspended Solids and Sediment

Species		General Tolerance	Preference for Turbid Systems
Common name	Scientific name		
Sturgeons			
Pallid sturgeon	*Scaphirhynchus albus*	X	X
Herrings			
Gizzard shad	*Dorosoma cepedianum*		X
Mooneyes			
Goldeye	*Hiodon alosoides*	X	
Carps and minnows			
Goldfish	*Carassius auratus*	X	
Lake chub	*Couesius plumbeus*	X	
Common carp	*Cyprinus carpio*		X
Silverjaw minnow	*Ericymba buccata*	X	X
Sturgeon chub	*Hybopsis gelida*	X	
Flathead chub	*Hybopsis gracilis*	X	
Bigmouth shiner	*Notropis dorsalis*		X
Red shiner	*Notropis lutrensis*		X
Sacremento blackfish	*Orthodon microlepidotus*	X	
Suckermouth minnow	*Phenacobius mirabilis*	X	
Mountain redbelly dace	*Phoxinus oreas*	X	
Fathead minnow	*Pimephales promelas*	X	X
Bullhead minnow	*Pimephales vigilax*	X	
Woundfin	*Plagopterus argentissimus*	X	
Creek chub	*Semotilus atromaculatus*	X	X
Suckers			
White sucker	*Catostomus commersoni*	X	
Bigmouth buffalo	*Ictiobus cyprinellus*	X	
Golden redhorse	*Moxostoma erythrurum*	X	
Bullhead catfishes			
White catfish	*Ictalurus catus*	X	
Black bullhead	*Ictalurus melas*	X	X
Pirate perches			
Pirate perch	*Aphredoderus sayanus*	X	
Sunfishes			
Green sunfish	*Lepomis cyanellus*	X	
Orangespotted sunfish	*Lepomis humilis*	X	
Redear sunfish	*Lepomis microlophus*	X	
Spotted bass	*Micropterus punctulatus*	X	
Guadelupe bass	*Micropterus treculi*	X	
White crappie	*Pomoxis annularis*	X	
Black crappie	*Pomoxis nigromaculatus*	X	
Perches			
Slough darter	*Etheostoma gracile*	X	
Least darter	*Etheostoma microperca*	X	
Johnny darter	*Etheostoma nigrum*	X	
Orangethroat darter	*Etheostoma spectabile*	X	
Sauger	*Stizostedion canadense*	X	
Drums			
Freshwater drum	*Aplodinotus grunniens*	X	

Source: Robert J. Muncy et al., *Effects of Suspended Solids and Sediment on Reproduction and Early Life of Warmwater Fishes: A Review,* prepared for U.S. Environmental Protection Agency, EPA 600/3¢79¢042 (Washington, D.C.: U.S. Government Printing Office, 1979), table 7, p. 72.

sedimentation as lower fish populations. And silty water can reduce the visual attractiveness of a recreation facility to picnickers, hikers, and others for whom the water serves to complement their primary recreational activity. One survey, taken in 1971, indicated widespread concern among recreational water site users about the impacts of sediment and other pollutants on their recreational activities (figure 3.3).

Swimming and Boating Accidents

High turbidity probably reduces the pleasure of swimming and boating, although there have apparently been no studies of how significant these impacts are. Excessive turbidity and sedimentation, by obscuring submerged hazards, may also be a cause of swimming, diving, and boating accidents. For instance, swimmers may become entangled in weeds growing in areas made shallower by sediment. High turbidity also is likely to make it more difficult to find and rescue drowning victims.

Even though no direct information exists on the number of accidents caused by excessive turbidity, there are data on the causes of recreational accidents that can provide an indication of how serious a problem turbidity might be.

In 1978, there were over 7,000 drownings in the United States.[56] More detailed earlier data indicate that, in 1971, almost half of the drownings were in rivers or small lakes, ponds, or reservoirs, and another quarter occurred in streams, farm ponds, canals, or larger lakes.[57] Turbidity may have been a factor in a large number of these drownings, but many undoubtedly were caused by factors unrelated to turbidity (for instance, a swimmer being swept out to sea, someone falling from a boat). Any estimate of the number of sediment-related drownings would be arbitrary, but if it were only 1 or 2 percent of the total, that would still account for approximately 100 deaths annually.

Data on boating accidents indicate that there were 6,573 vessels involved in accidents in 1981, resulting in 1,208 fatalities.[58] Most of the accidents resulted from collisions with other vessels, but most of the deaths resulted either from capsizing or passengers falling overboard. Nine-hundred thirty vessels grounded or collided with fixed objects, causing about 100 deaths; many of these accidents could have been associated with turbidity and sedimentation if murky water obscured the water body's depth or the location of a fixed object.

Another possible cause of boating accidents associated with sedimentation could be the ability of shallow lakes to "kick up" faster in a wind and to produce choppier waves. Although no examples have been

Figure 3.3
Possible Harmful Effects of Pollutants on Recreational Activities

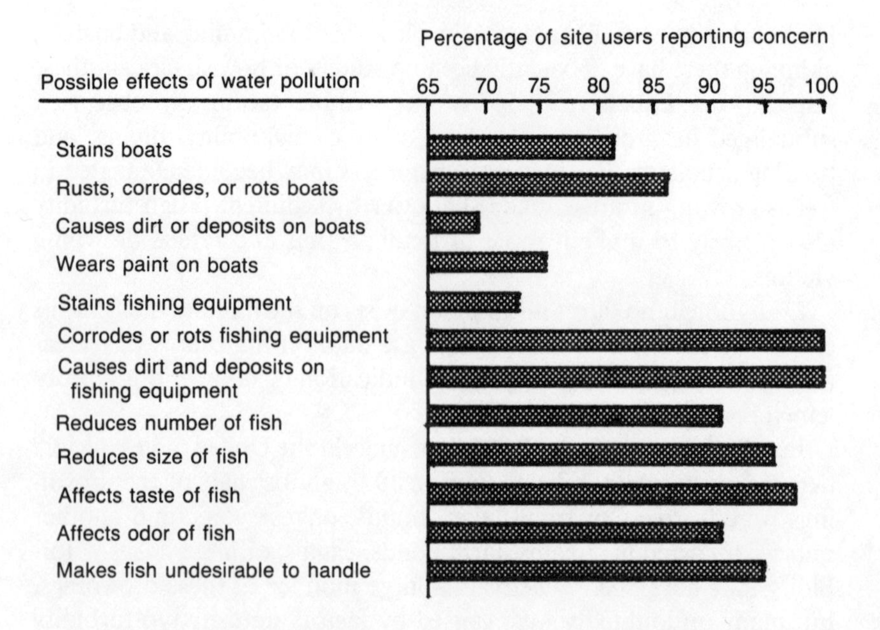

Source: Task Force on Agriculture Nonpoint Sources of Pollution, *Final Report* (Springfield, Ill.: Illinois Institute for Environmental Quality, 1978), table 8, p. 49.

found documenting this type of effect, excessive sedimentation certainly could reduce the depths of lakes sufficiently to make boats more likely to capsize.

A reasonable estimate, based on the available data, is that as many as 200 fatalities a year could be associated with excess turbidity. Nonfatal injuries and other damages that might be associated with turbidity or sedimentation undoubtedly are much higher. For instance, among swimmers, there are about 50 injuries for every fatality.[59]

Recreational Fishing

Turbidity and sedimentation can significantly decrease the quality of sport fishing. This is not just because they reduce fish populations and cause less desirable fish—for instance, carp—to replace more highly valued game fish such as trout in a body of water.[60] The fish that do exist in sediment-laden waters are harder to catch; turbidity directly reduces the angler's chance of success by decreasing the distance at which the fish can see a lure.[61] The data from a six-month study conducted on an Illinois lake indicated that the percentage reduction in the rate at which fish are caught is approximately proportional to the percentage reduction in visibility (figure 3.4). The author of a study of several ponds and reservoirs in Oklahoma concluded, "The clear reservoir attracted more anglers, yielded greater returns per unit fishing effort, as well as more desirable species, and was immeasurably more appealing in the aesthetic sense."[62]

Recreational Facilities

Recreational facilities also can be seriously affected by sediment in a water body. Good fishing holes can be so silted up as to be useless. Swimming beaches can become too shallow to be enjoyable or can have their bottoms so covered with slimy muck that no one wants to use the beaches. Docks and piers constructed to serve recreational boaters may be rendered useless because sedimentation prevents boats from reaching them. In 1976, an Illinois historical survey crew found it impossible to float a canoe in more than one-third of the surface area of several bottomland lakes it was surveying.[63] And the annual speedboat races and regattas that were held on Lake DePue in Northern Illinois had to be terminated in 1974 because the lake had become too shallow.[64]

Aesthetics

Most water-related recreational activities do not, in fact, involve swimming, boating, or fishing. Rather, they involve people picnicking, walk-

Figure 3.4
Effect of Turbidity on Fishing Success

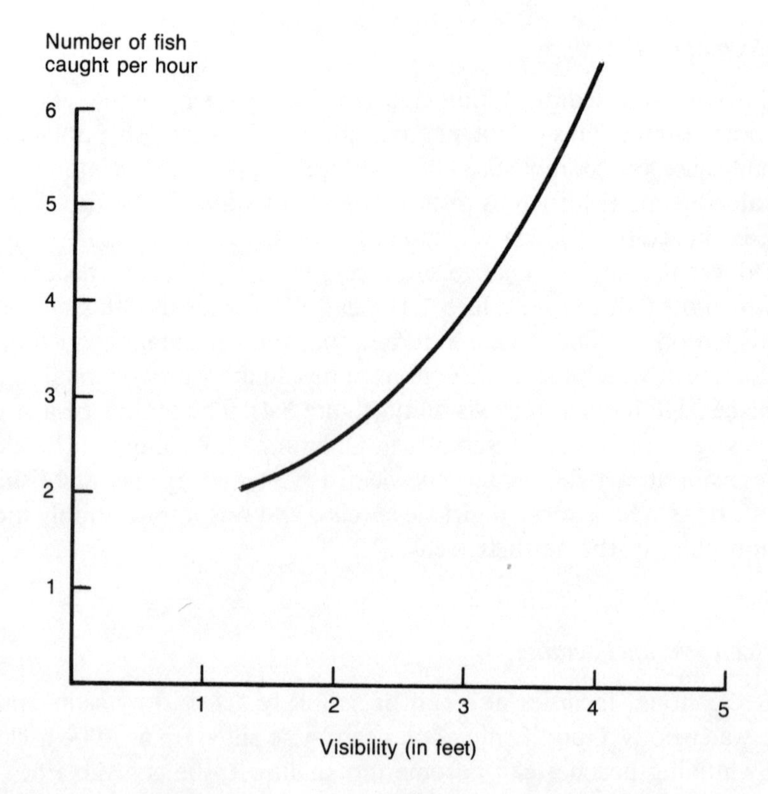

Source: Richard E. Sparks, "Effects of Sediment on Aquatic Life" (Report to the Illinois Task Force on Agriculture Non-Point Sources of Pollution, Subcommittee on Soil Erosion and Sedimentation, June 18, 1977), p. 7.

ing, or playing near the water. In such cases, turbidity and sedimentation cause less direct harm than in the previous cases, but they do nevertheless diminish the quality of the recreational experience. Moreover, the aggregate impact of less attractive water bodies may be very high because so many people are affected.*

---

*A study of the value of different types of recreational experiences along the Poudre River in Colorado, for example, generally found that individual anglers and white-water rafters placed a higher value on maintaining adequate amounts of water in the river than did people participating in shoreline recreation. However, because two to three times as many people were engaged in shoreline recreation, the aggregate value they placed on the resource was almost as high as the other two types of recreationists combined. Although this study pertained to water quantity rather than water quality, its conclusions can reasonably be generalized.[65]

Americans clearly want their waters to be clean. A Louis Harris poll in 1982 reported that 74 percent of all Americans rated curbing water pollution as very important.[66] This attitude partially reflects a fear of the health effects that can result from toxic contaminants and other pollutants. But it also undoubtedly reflects a basic desire for clean rather than turbid water.

Water Storage in Lakes and Reservoirs

Sediment affects lakes and reservoirs by reducing their water-storage capacity, changing the temperature of the water, and providing increased opportunities for the growth of water-consuming plants.

The filling of lakes by sediment is an inevitable process that has continued since the Earth was first formed. Nevertheless, this siltation can be costly, particularly when it is accelerated by excessive erosion and when it involves reservoirs created by constructing expensive dams.[67] Reservoirs supply valuable water to U.S. farms, municipalities, and industries, provide recreational opportunities for millions of visitors a year, and protect downstream areas from flooding. All of these benefits can be diminished by sedimentation, and building replacement facilities may prove very expensive.

For most U.S. reservoirs, sedimentation is not a major problem. The total capacity of reservoirs in the United States was estimated in 1978 to be approximately 225 trillion gallons, or 690 million acre-feet.[68] The most extensive study of reservoir sedimentation rates in the United States found that slightly more than 0.2 percent of the reservoir's initial capacity was being filled by sediment each year.[69] This implies an average of 1.4 to 1.5 million acre-feet of capacity consumed by sedimentation annually.* Another investigator estimated that, nationwide, 1 million acre-feet of reservoir capacity are filled with sediment each year.[71]

It may be hundreds of years before sediment deposits fill even the storage area reserved for this purpose in some of the large Tennessee Valley Authority reservoirs.[72] And many U.S. Bureau of Reclamation reservoirs are "silting up" more slowly than expected.

However, many reservoirs are continuing to lose capacity at a rapid rate:

- A nationwide study in 1968 of 968 reservoirs found that 15 percent were losing storage capacity at rates exceeding 3 percent a

*A report published by the U.S. Geological Survey has a lower estimate of reservoir capacity—about 500 million acre-feet, which would give an annual sedimentation loss of about 1 million acre-feet annually.[70]

year; for 2 percent of the reservoirs, the rates exceeded 10 percent a year.[73]

- A survey conducted in the late 1940s concluded that water-supply reservoirs in midwestern states were silting up so rapidly that one-third of them would have to be replaced within 50 years.[74]
- A survey of hydroelectric reservoirs in the Piedmont found that 13 of them had completely silted up within an average of 30 years of their construction.[75]
- A reservoir on the New River, Virginia, lost 80 percent of its capacity within 18 years.[76]
- The capacity of the small reservoir behind Imperial Dam on the lower Colorado River declined by 71 percent within 5 years of its construction and by 81 percent within 6 years.[77]

In some cases, the capacity loss can be especially rapid. During the 1938 flood, some California reservoirs lost over one-third of their capacity.[78] And Lake Harding, also in California, was nearly filled within a month after fires denuded its watershed in 1926.[79]

As chapter 2 explained, smaller reservoirs tend to have higher rates of capacity loss than larger reservoirs (figure 3.5). Mauldin's milldam

Figure 3.5
Sedimentation in U.S. Reservoirs

Reservoir capacity (acre-feet)	Number of reservoirs	Total initial storage capacity (acre-feet)	Total storage depletion (acre-feet)	Total storage depletion %	Individual reservoir storage depletion Average (%/yr.)	Individual reservoir storage depletion Median (%/yr.)	Average period of record (years)
0–10	161	685	180	26.3	3.41	2.20	11.0
10–100	228	8,199	1,711	20.9	3.17	1.32	14.7
100–1,000	251	97,044	16,224	16.7	1.02	.61	23.6
1,000–10,000	155	488,374	51,096	10.5	.78	.50	20.5
10,000–100,000	99	4,213,330	368,786	8.8	.45	.26	21.4
100,000–1,000,000	56	18,269,832	634,247	3.5	.26	.13	16.9
Over 1,000,000	18	38,161,556	1,338,222	3.5	.16	.10	17.1
Total or average	**968**	**61,239,020**	**2,410,466**	**3.9**	**1.77**	**.72**	**18.2[1]**

[1] The capacity-weighted period of record for all reservoirs was 16.1 years.

Source: Ferris E. Dendy, "Sedimentation in the Nation's Reservoirs," *Journal of Soil and Water Conservation* 23 (1968):137.

on a stream in Hall County, Georgia, provides an interesting example of this phenomenon.[80] The dam was built around 1865, but, by the early 1900s, the eroding hills had filled the pond with sediment. The aggradation continued until about 1930, when the waterbed was about 16 feet above its original level and the surface of the water itself was 4 feet above the top of the dam. The stream's banks also had risen as sediment deposited along them. The amount of sediment entering the stream has since decreased because of reduced erosion in the watershed. As a result, the stream began to erode by about 1955, and, by the mid-1970s, the bed had fallen 7 feet, uncovering 3 feet of the dam.

The author of one of the most extensive analyses of sedimentation in reservoirs concluded, "If present siltation rates continue, about 20 percent of the Nation's small reservoirs will be half filled with sediment and, in many instances, their utility seriously impaired in about 30 years."[81]

Substantial geographical variations also exist in lake and reservoir sedimentation rates. Areas with naturally high erosion rates experience the fastest rates of storage depletion. In a study of 42 reservoirs in Iowa, Nebraska, and Missouri, 18 lost more than 25 percent of their storage capacity in 11 years or less.[82] This area always has been particularly prone to erosion because of its topography, the nature of its soil, and its intense rainstorms. On the other hand, areas with relatively low erosion rates, such as Pennsylvania and Maryland, experience lower sedimentation rates (figure 3.6).[83]

Sediment also may affect the ability of lakes and reservoirs to provide an adequate water supply by changing the rate at which water is lost through evaporation or transpiration. In this case, however, the sediment has both a beneficial and a detrimental effect, and it is not clear which is more important.

The beneficial effect results from the impact of sediment on water's absorption of solar energy. As mentioned earlier, sediment-caused turbidity prevents most of the sunlight from penetrating much below a water body's surface. This causes a sharp temperature stratification, with a thin layer of warm water on the surface and much cooler water below. Since the temperature of the surface rather than the average temperature of a water body controls the evaporation rate, this effect, by itself, would increase evaporation from the reservoir. However, the turbidity has an additional effect—to reflect the sunlight back into the atmosphere so that it is not absorbed by the water at all. This results in the water on the surface as well as below being cooler and, therefore, evaporating more slowly than it would otherwise.

Both theoretical calculations and empirical measurements indicate

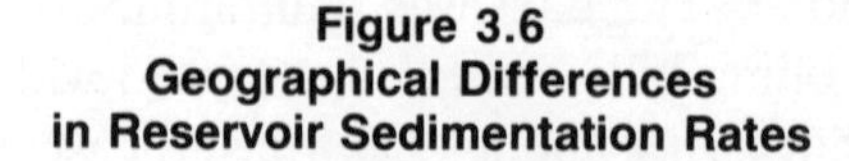

Figure 3.6
Geographical Differences in Reservoir Sedimentation Rates

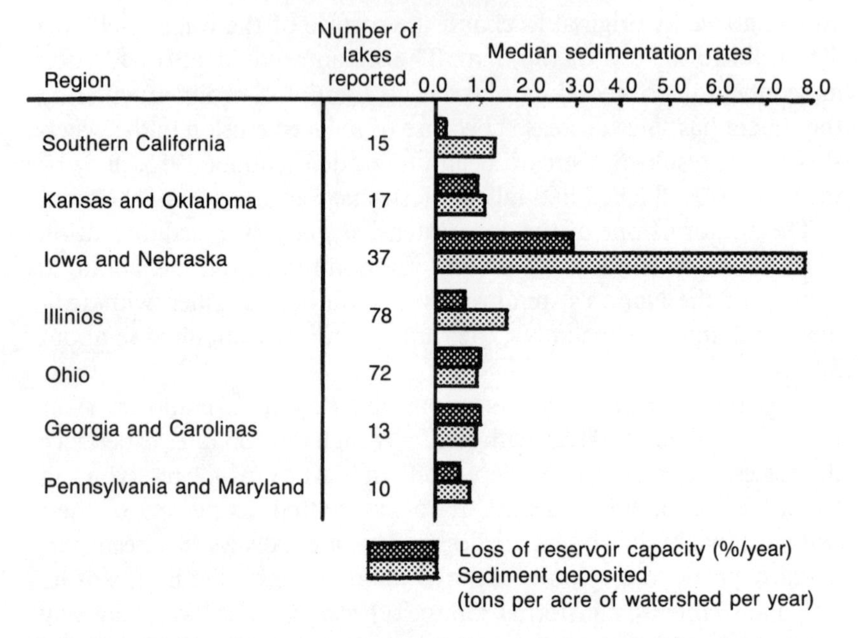

Source: John B. Stall, "Soil Conservation Can Reduce Reservoir Sedimentation," *Public Works*, September 1962, table 1, p. 125.

that turbidity's cooling effect is more powerful than its stratification effect.[84] One study measured water temperature in two adjacent reservoirs, one turbid and the other clear (figure 3.7), and found that the water temperature in the turbid pond not only fell sharply below the surface but also was lower at the surface itself. Thus, the net effect of the suspended sediment would be to reduce the rate of surface evaporation.

This effect could be important in the western United States, where water often is stored for several years so that it is available during a drought. The more evaporation that occurs, the greater the amount of water that must be stored to maintain sufficient reserves for a shortage. The cost of this storage can be quite significant (see chapter 5). However, the beneficial effects of turbidity are probably very small. Most of the reservoirs there are so large that suspended sediment settles out of the water near the inflow, leaving the majority of the surface water clear. There have, however, been no analyses of the importance of this evaporation phenomenon.

Figure 3.7
Water Temperature Profiles for Two North Mississippi Reservoirs

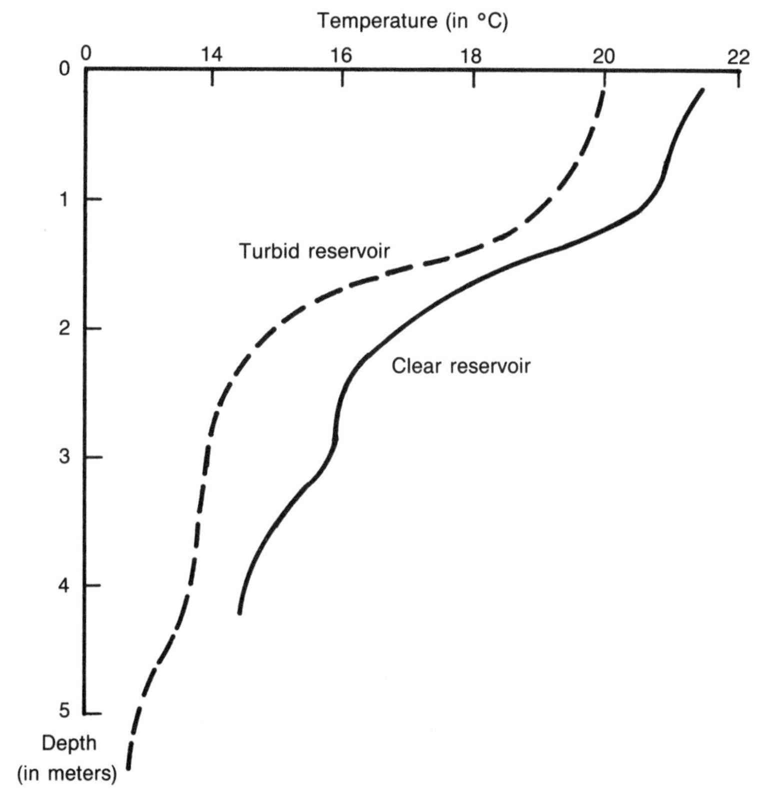

Source: Frank R. Schiebe et al., *Suspended Sediment, Solar Radiation and Heat in Agricultural Reservoirs,* Proceedings of the Third Federal Inter-Agency Sedimentation Conference, Denver, March 22–25, 1976 (Washington, D.C.: Water Resources Council, 1976), pp. 3–10.

Moreover, whatever positive impact suspended sediment has on evaporation may be offset by various negative effects resulting from sedimentation. Since sedimentation reduces the average depth of a reservoir without changing the surface area it covers, the reservoir absorbs the same amount of solar energy but contains a smaller volume of water. As a result, unless other factors intervene,* the average water temperature is higher, which should increase evaporation.

*The fact that the water-retention time is reduced in a shallow reservoir at least partially offsets this effect since the water has less opportunity to be warmed before it leaves the reservoir. If the reservoir experiences a constant rate of inflow and outflow during the year, these two effects could be expected to exactly offset one another.

In addition, to compensate for lost capacity caused by sedimentation, a reservoir operator may raise the level of the water surface. This, however, will increase the surface area and, therefore, the amount of solar energy absorbed by the water. Again, the result will be increased evaporation. At the Elephant Butte, New Mexico, and Roosevelt, Arizona, reservoirs, the annual evaporation was estimated to have increased by 1,500 and 2,500 acre-feet, respectively.[85]

Finally, sediment deposits often provide a good site for plants that consume and transpire (that is, release to the atmosphere through their leaves) large quantities of water.[86] In the western United States, such plants are thought to cause significant water wastage, although the evidence on this is not clear.[87] Such vegetation, however, can also provide benefits—creating valuable wildlife habitat and functioning as a "sediment trap," decreasing the amount of sediment carried into a lake or reservoir.[88]

Navigation

The sedimentation occurring in harbors, bays, and navigation channels reduces the capacity of these facilities to handle commercial and recreational craft, increases the likelihood of shipping accidents, and requires expensive dredging to keep the facilities usable.

Total waterborne commerce in the United States amounts to 1.7 billion tons a year.[89] About two-thirds of this is domestic (that is, it both originates and terminates within the United States), accounting for 16 percent of annual total domestic transport by weight. Approximately 25,000 miles of channels, 107 commercial ports and harbors, and 400 small boat harbors serve this traffic. Most of these facilities are maintained by the U.S. Army Corps of Engineers.

Sedimentation has created navigation problems since settlers first began arriving in America. Joppa Town, for instance, was "the most important and prosperous seaport of Maryland" after it was founded in 1700.[90] It had an excellent location, at the head of a large deep bay, suitable for the largest oceangoing ship and convenient to a rich farming area. This farmland played a major role in the port's prosperity—and its destruction. Within 50 years of Joppa Town's founding as a busy port exporting tobacco and other commodities, so much sediment had filled the bay that an increasing amount of traffic was transferred to Baltimore. Today, even at high tide, the water is a mile and a half from the site of the town and the scene of Joppa Town is one of desolation. Old foundations are still visible through the tangled growth of weeds and underbrush. Between 20 and 30 feet inland from the original shoreline is a heap of stones, the remnants of the old wharf.

A hundred feet further in is tree-covered land where ships once rode at anchor.

Joppa Town's experience is an extreme case, but sediment continues to cause serious navigational problems in the United States. The major problem, maintaining adequate channel depths, has become more acute in recent years as vessels have become larger. The most common response is to dredge sediment out of the channels and harbors where it accumulates. About 500 million cubic yards of material are dredged from these sites each year, 50 to 60 percent of it by the Corps of Engineers.[91] This is over twice the volume of material (211 million cubic yards) excavated in creating the Panama Canal.[92] Even so, it is not enough to keep the facilities clear.

Dredging can create problems of its own.[93] Dredged material may contain significant amounts of heavy metals and other toxic contaminants. Stirring up a waterway's bottom deposits by dredging may release these contaminants, making them available for uptake by fish and other aquatic wildlife. If the dredge spoils are deposited back in the water away from the channel, this effect is compounded, and, in addition, many of the other problems associated with sedimentation can be caused again. Depositing the spoils on land also may be unsatisfactory, since they could destroy valuable wetlands and create health hazards.

Trying to avoid these problems can be very expensive. Disposing of materials dredged from the Illinois River costs almost as much as the dredging itself.[94] The Corps of Engineers estimates that avoiding environmental impacts may increase the costs of disposing of dredged material 15-fold.[95] Often, attempting to find a suitable disposal method also has significantly delayed channel maintenance efforts.[96]

Sometimes, however, proper disposal of dredged material can create benefits. For instance, a proposed project to improve access to the ports of New Orleans and Baton Rouge would use the dredged material to create additional wetlands along the coast, partially offsetting the losses that occur elsewhere in that region.[97] In other cases, the dredged material has been used to replenish beaches and stabilize coastal areas.[98]

Dredging is not always enough, so expensive "river-training" projects such as jetties have to be constructed as supplements to the dredging operations. Texas's Colorado River was depositing so much sediment in the Gulf Intercoastal Waterway where the river and waterway intersect that locks had to be constructed that would exclude sediment but allow ships to pass through.[99] And engineers have been constructing increasingly complex and expensive channel works along the lower Mississippi River for close to 200 years to maintain adequate channel depths.[100] Even so, ships attempting to reach New Orleans, Baton Rouge, or other nearby harbors often experience extensive delays and increased risks of accidents because of shoaling at the river's mouth.[101]

Other In-Stream Impacts

Several other in-stream effects have been identified, but apparently none has been documented or analyzed extensively. One such effect is the increased wear that may be experienced by hydroelectric and other machinery that operates in the water.[102] Water flows through hydroelectric turbines with enough force that even pure water can cause serious damage to the metal blades and casings. The addition of sediment to the water may increase its abrasive power and, as a result, the damage it causes.* The loss of reservoir storage capacity can also adversely affect hydroelectric installations.

The abrasive properties of suspended sediment could also damage a ship's propellers (and other equipment that operates at high speeds in water) and interfere with the efficient operation of cooling systems for marine motors. However, no documentation of such problems has been found.

OFF-STREAM IMPACTS

In 1980, an estimated 290 billion gallons of water a day were withdrawn from U.S. surface waters for all uses (including agricultural, industrial, and municipal).[104] The addition of suspended sediment to a given volume of water increases both the water's volume (since the sediment displaces some of the water) and its density (since the sediment is heavier than the water it displaces). Both factors may affect the cost of pumping water to make use of it off-stream. The increase in volume may also increase the magnitude of floods.

Sedimentation in drainage ditches, before runoff even reaches a waterway, can result in flooding and increase ditch maintenance costs. Similarly, sedimentation in pipes and canals carrying water from a river can reduce the amount of water available to users and increase the

---

*A potential, partially compensating effect may result from the sediment increasing the weight and volume of water flowing through the turbines. However, this effect is probably very small. If the water flowing through a dam contains 800 milligrams of suspended sediment per liter (approximately the mean value in U.S. rivers), it would cause a weight increase of only 0.08 percent and a volume increase of 0.04 percent.[103] Because most hydropower is generated by water stored in a reservoir, the benefits would be much less than these figures indicate.

operation and maintenance costs of getting it to the points of use.

Finally, sediment increases the cost of treating water before it is used and may cause health and taste problems with drinking water. If turbid water is not treated prior to its use, the suspended particles can increase wear on the machinery and equipment through which it passes, decrease the efficiency with which this equipment operates, and increase the amount of maintenance that must be carried out on the equipment.

Flood Damages

Over 100 million people and 15,000 communities and recreation areas are in flood-prone areas in the United States.[105] Even if soil erosion were eliminated, flooding would remain a problem. Nevertheless, in several different ways, the process of erosion and sedimentation is making a serious problem worse.*

The changes that sedimentation can cause in a stream channel have several different effects on flood damages. In any stretch where a streambed has aggraded, the elevation of the water surface associated with any volume of flow will be higher and may flood adjacent land in the absence of what would otherwise be a flood flow. For instance, Needles, California, grew up on the banks of the then free-flowing lower Colorado River. Between 1935 and 1950, however, the construction of three dams—Hoover, Parker, and Davis—radically changed the river's character. The most serious changes, as far as Needles was concerned, came with the completion of Parker Dam and the filling of Lake Havasu, which begins 25 to 30 miles downstream of the town. The creation of the lake both raised water levels upstream of the lake and slowed the water. A contemporary observer described the results:

> The reduced velocities of flow caused the silt to be dropped, forming extensive sand bars in the former river channel. When the sand bars neared the water surface, tules and willows started to grow, forming an almost impenetrable jungle across the former river channel and river bottom lands.

---

*An important, though indirect, relationship between sediment and flooding is not addressed here. This is the fact that many of the factors that increase soil erosion also increase storm runoff, as indicated in chapter 2. The vegetation and other factors that reduce soil erosion usually will also slow the rate at which water runs off land, thus allowing more of it to seep into the ground. With eroding fields, not only is the amount of runoff greater, but also the water runoff is faster. Consequently, more storm water is delivered to a river in a shorter period of time, increasing the peak flow. Both effects—the total volume of runoff and the size of the peak flow—increase flood damages. Neither effect is caused directly by erosion, but both are associated with it.

> Through this mass of vegetation, the water would filter, dropping its silt load as it went. The result was a further rise of water surface upstream, further formation of sand bars, new growth of tules and willows, and so on.
>
> By 1942, the process had reached half way from Topock to Needles, and by late 1944, it had reached the town of Needles itself. The resultant rise of water surface at Needles has been approximately 1.5 meters (5 feet).[106]

The increasing innundation was finally halted in 1951 by cutting and maintaining a new river channel through the accumulated silt deposits. But it had forced the abandonment of nearly 100 dwellings and threatened the main line of the Santa Fe railroad.*

The history of the small village of Soldiers Grove, in southwestern Wisconsin, provides a similar example.[107] Built in the floodplain of the normally docile Kickapoo River, the village was rarely bothered by flooding until early in this century when sediment from the surrounding hills began to clog the river channel and raise its bed. Disastrous floods began to buffet the village with increasing frequency, resulting in proposals for costly flood protection works. Finally, however, following a 1978 flood, Soldiers Grove, with support from the U.S. Department of Housing and Urban Development, decided to completely relocate onto higher ground, at a cost of over $6 million.

These are by no means the only examples of aggrading rivers flooding adjacent land, creating swamps and stagnant bayous, and interfering with normal drainage of riparian lands.

Such conditions destroy land's ability to grow crops. Even when riparian land is in a wildlife reserve (for instance, along the scenic Hatchie River in Tennessee), such swamping can adversely affect wildlife habitat.[108] And, not very long ago, it could lead to serious outbreaks of malaria.[109]

Even if sediment causes no permanent inundation of riparian lands, the aggradation can cause serious periodic flooding in stretches where flooding previously was rare. Examples of such flooding problems have been observed from California to North Carolina. However, to partially offset the flooding caused by the aggradation, the heaviest sediment in the floodway may settle out as soon as the water leaves the channel, causing berms to build up along each bank. Such berms, however, can be breached during heavy floods, allowing the full volume of the waterway to pour across its floodplain.

---

*This would have been true even if the slope of the stream channel had remained unchanged. However, since aggradation usually results in a flattening of the slope, water flows more slowly. As a result, water depth must increase for it to pass a particular volume of water.

Sedimentation also makes flood damage two to three times greater along the upstream stretches of a watershed than along a major river or its principal tributaries.[110] This is partly due to the decreased flood-carrying capacity of silt-clogged streams. However, these increased damages along upstream reaches where sedimentation is occurring may be offset at least partially by decreased flooding downstream. The reduced capacity of a stream in an aggraded reach may reduce the size of the peak water flow downstream and delay its progress.*

In addition, sedimentation can decrease a flood-control reservoir's ability to protect downstream areas. In a multipurpose reservoir, the flood-control storage is provided at the top of the reservoir, which is the last to be affected by sedimentation (figure 3.8). However, a single-purpose flood-control reservoir, which often may be dry except when

Figure 3.8
Typical Allocation of Reservoir Storage Capacity

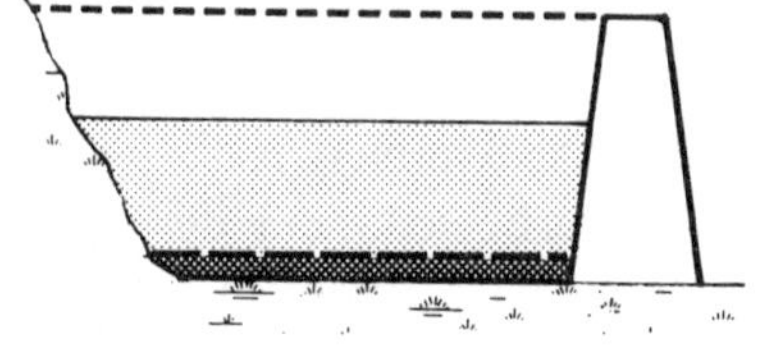

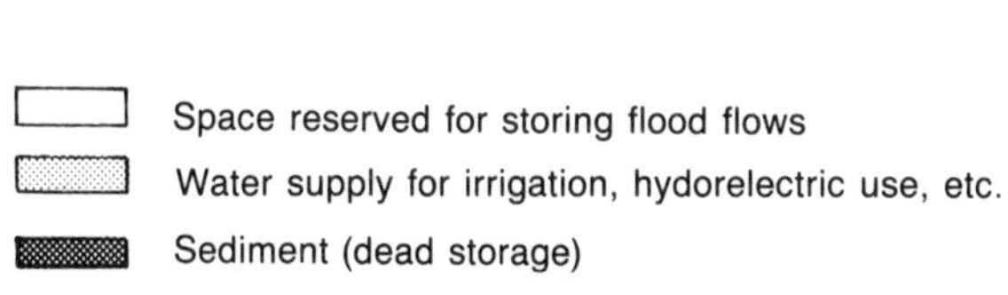

Source: U.S. Army Corps of Engineers, *Washington Metropolitan Area Water Supply Study, Final Report* (Baltimore: U.S. Army Corps of Engineers, 1983), p. 47.

*Sometimes, however, delaying the peak flow on a tributary may cause it to coincide with and thus add to the peak flow in the main river rather than to precede it. In such cases, the increased flooding on the tributary will cause increased flooding on the main river as well.

flood flows occur, may be partially filled with sediment by only one or two storms.[111] In such cases, the reservoir must be excavated if the people downstream are not to lose the protection it was built to provide.

Another way in which erosion can affect flood damages results from the fact that large volumes of suspended sediment increase the volume of water (or, more precisely, the volume of the water-sediment mixture) in a stream or river and therefore increase the volume of floodwater (figure 3.9). With a suspended sediment concentration of 100,000 parts per million (or 10 percent), water volume increases by about 4 percent. As indicated in chapter 2, the suspended sediment concentration can, in unusual cases, be as much as five times this amount.

More serious flood damages occur, however, when sediment settles out of water on a floodplain. Although in some cases such deposits can be beneficial,* on balance they do more harm than good. Deposits of sterile silt, sand, and gravel have covered fertile riparian agricultural lands in many areas of the United States—among them, California, Vermont, Utah, Georgia, and Texas.[114] In the eastern states, according to one report, "Large areas of once rich and fertile bottom land . . . have been buried under heavy deposits of infertile soil materials."[115] Ironically, the silt deposits, together with increased flooding, have forced the bottomland farmers into the uplands, where their land clearing has caused more erosion and more downstream damage.[116] In such cases, even if the sediment is relatively fertile, it can smother the current year's crops.[117] Similarly, sediment deposits on rangeland can smother grasses and otherwise diminish its suitability for raising cattle.[118]

Finally, serious problems also are caused when floods deposit sediment in urbanized areas. As one observer commented, "Unless one has attempted to clear the sediment from homes and businesses follow-

---

*Before the construction of the Aswan Dam in Egypt, farmers along the Lower Nile used to welcome the fertile soils deposited on their fields by the annual spring floods.[112] Similarly, the construction of levees along the lower Mississippi River has eliminated the natural deposition of sediment on nearby wetlands. These wetlands, which provide valuable wildlife habitats, are formed of alluvial deposits that are continually consolidating. The annual sediment deposition is necessary to offset the consolidation and keep the land surface above the water. With that deposition eliminated, continued consolidation has caused the lands to sink below the water level, killing the vegetation growing there. Louisiana is losing such lands at a rate of 16 square miles per year.[113]

Figure 3.9
Effect of Suspended Sediment on Weight and Volume of Water

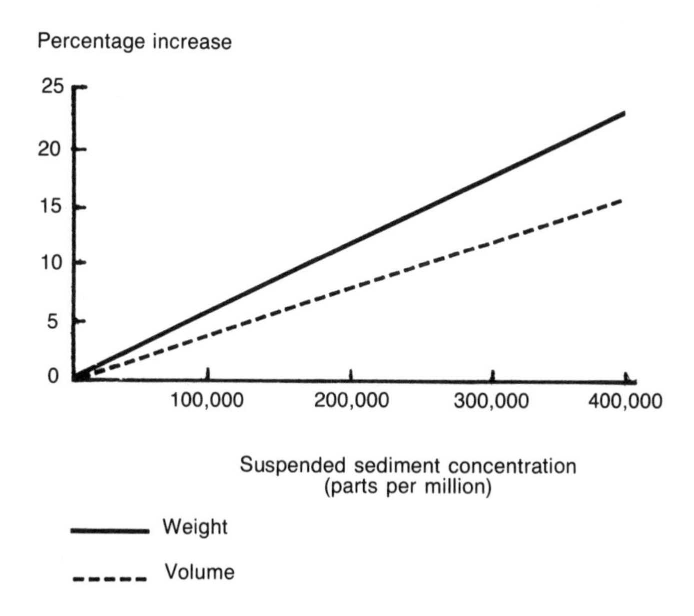

ing a flood, it is difficult to realize the extent of the problem."[119] Another, after inspecting flood damage records, concluded that "a relatively high proportion of the flood damages in urban areas are due to the cost of cleaning sediment from streets, houses, furniture, etc."[120]

Water Conveyance

Sediment-laden water also can have serious effects as it flows through water-conveyance systems, either on its way to a waterway or after it has been removed for an off-stream use. In addition, high turbidity can increase the cost of pumping water from its source.

Sedimentation in drainage ditches can increase flooding along and upstream of the ditches and will require increased maintenance efforts to remove the sediment.* Studies from Illinois and Ohio indicate that this can be a very significant problem.[121] In Illinois, a survey of the State Highway Department maintenance districts and county maintenance operations found that they were removing 1.1 million tons of sediment annually.[122]† The study estimated that statewide a total of 2.5 million cubic yards of sediment—1.4 percent of the Illinois's total erosion—were removed from ditches each year. However, these data probably understate the problem; several respondents noted that severe budget constraints prevented them from performing the regular ditch cleaning that was needed near cropland and housing developments.

Similar problems are also encountered with canals, channels, and aqueducts used to carry water to its final point of use. This includes the approximately 110,000 miles of irrigation channels in the United States.[123] A study of the Delta-Mendota Canal in California found that it has had frequent capacity problems since the 1960s because of sedimentation.[124] To keep these problems manageable, the canal is drained for repairs and sediment removal for 8 to 10 weeks roughly every two years. Because of the reduced carrying capacity and the closure times required for maintenance, delivery shortfalls of 400,000 to 450,000 acre-feet of water per year are forecast for the San Joaquin Valley irrigation districts.

Fortunately, most water-conveyance structures are designed to ensure that the water in them moves fast enough to prevent serious sedimentation from occurring. Thus, such problems should not be extensive unless (*a*) there has been a significant increase in the amount of sediment or the size of the sediment particles entering such a conveyance facility, or (*b*) the facility is in an area (such as an irrigation channel in a field) where topography and economic considerations prevent appropriate design standards from being adopted.

An additional impact of sedimentation on water conveyance is its ability to increase the cost of pumping water. These effects may never be noticed on an isolated basis, given the miminal effect that suspended sediment has on water weight and volume. A crude estimate of national impacts, however, suggests that pumping the suspended sediment could require an additional 5 to 6 million kilowatt hours of elec-

---

*Parallel problems may occur with storm sewers, but, since they are primarily affected by erosion in urban areas, they are not dealt with here.

†The survey covered 8 or 9 state highway maintenance districts and 44 of 102 counties.

tricity a year.[125]*

Suspended sediment may also affect the cost of water conveyance by increasing the wear on pumps. However, it is not possible with existing information to estimate the total pump wear attributable to suspended solids in water. In areas of the United States where this could be a serious problem, coarse filters or small settling basins can be installed to remove most of the harmful particles prior to pumping.

Increased Water-Treatment Needs

In 1980, public water supplies withdrew an average of 22 billion gallons of water a day, and industries (other than thermoelectric power) withdrew directly another 29 billion gallons per day of fresh water from surface-water sources.[126] Much of this industrial water, however, is used for cooling and other purposes where high-quality water is not necessary. Still, even though the majority of industrial water use does not need to be high quality, much of it may nevertheless require treatment by filtration or sedimentation to remove suspended solids to prevent hydraulic machinery from being clogged.

Drinking water, by contrast, must be essentially sediment-free to ensure that it does not cause damages when used for drinking or washing, that it is pleasing to the user, and that turbidity does not interfere with attempts to kill bacteria and viruses or to remove chemical contaminants from water supply.[127] To achieve EPA's required level of purity, a water-supply company must treat its water to remove sediment.[128] A typical treatment process involves first having the water sit in sedimentation ponds so that larger sediment particles settle out. Chemical coagulants are often added at this stage to hasten the sedimentation, particularly for smaller particles. Sedimentation is followed by filtration, usually through a filter made of sand. Next, the water usually is treated with a disinfectant, such as chlorine, to kill any remaining microorganisms and may be chemically treated to remove other contaminants.

As the turbidity of a water supply increases, so do both the investment and the operation and maintenance costs of the water-treatment facility. At a very low level of turbidity, coagulation and large settling tanks may not be needed at all.[129] A treatment facility for highly turbid water, however, can be quite expensive to construct.[130] Increased

---

*This calculation assumes that the average pumping height in the nation is 100 feet, that average pumping efficiency is 60 percent, and that the average sediment concentration is 100 milligrams per liter.

turbidity also means that the filters have to be cleaned more frequently (figure 3.10). In addition, higher turbidity levels that remain after treatment require the use of greater amounts of chlorine or other disinfectants to purify the water.[131]

Municipal water systems attempting to deal with increased turbidity levels also have experienced serious taste and odor problems, often associated with the high chlorine dose required to ensure that bacterial levels are brought down to a satisfactory level.[132]

Moreover, the use of chlorine as a disinfectant may lead to additional health risks. If the water contains natural or synthetic organic substances (that is, compounds containing atoms of carbon), the higher chlorine doses may react with these compounds to create potentially toxic organic compounds such as chlorinated phenols and trihaleomethanes (for instance, chloroform, bromodichloromethane, dibromochloromethane, and bromoform).[133] Chloroform is a known animal carcinogen and a suspected human carcinogen. The carcinogenic properties of many of the other chlorinated compounds are unknown.

Figure 3.10
Impact of Sediment on Frequency of Filter Cleaning for Water-Treatment Plants

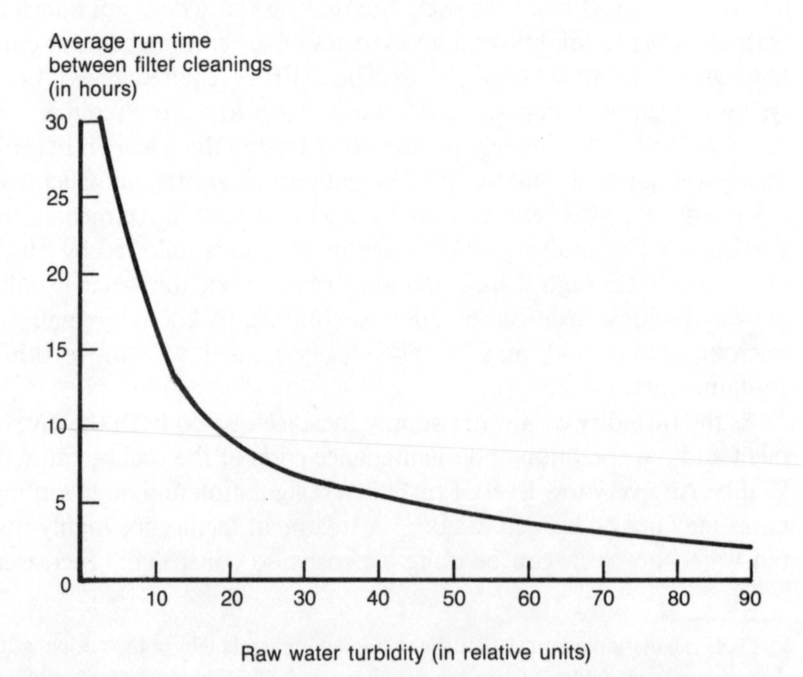

Source: Richard F. McCormick and Paul H. King, *Direct Filtration of Virginia Surface Waters: Feasibility and Costs*, Bulletin 129 (Blacksburg: Virginia Water Resources Research Center, 1980), table 21, p. 95.

Other Off-Stream Impacts

Sediment in water to be used for purposes other than drinking can also cause adverse effects. For instance, suspended sediment in irrigation water can form a crust on a field, reducing the amount of water that seeps into the soil, inhibiting the emergence of plants, and preventing adequate soil aeration.[134] A superintendant of the Imperial Irrigation District in California described some of the problems that can occur when silt seals the soil surface in irrigated fields.[135] For instance, with crops like alfalfa, evaporation from the standing water can "scald" the crop. With furrow-irrigated crops, such as lettuce and cantaloupes, the farmer may have to dry the field and break up the film of deposited silt before sufficient water can reach the root zone of the crops.

Such crusting can lead to further adverse effects. Because the irrigation water infiltrates less, increased amounts of water flow down the furrows, off the end of the field, and through the drainage ditches, potentially increasing erosion at each step. Because less water reaches the root zone and groundwater aquifer, the farmer must increase the amount of irrigation water taken from surface sources, compounding the other off-stream effects of suspended sediment and making the water unavailable for use by others. Evaporation from the water standing in the field also represents a waste of increasingly scarce and valuable water supplies.

In addition, the sediment may coat the leaves of young plants. The result is stymied plant growth (because photosynthesis is impeded) and reduced crop marketability (because the crops are both dirty and smaller).

Using untreated turbid water also can increase industrial expenses. Suspended sediment can increase the rate at which pumps and other equipment wear out, particularly in machinery having parts that fit together closely. One expert estimated that over 90 percent of hydraulic equipment failures can be traced to abrasive wear from silt and other particles.[136] Not all this expense stems from sediment in incoming water, however. Many firms attempt to avoid these abrasion problems by buying treated water from a municipal system or by treating it themselves before using it. For them, existing failures are largely caused by abrasive particles generated within the plant rather than by silt in the water supplies.[137]

Silt in untreated water may also settle in, or otherwise clog, cooling equipment.[138] This decreases the efficiency with which a cooling system operates, which, in turn, requires construction of a larger cooling facility, increases the operation and maintenance costs associated with the current facility, or reduces the efficiency of the entire production process of which the cooling system is a part. With a centrifugal water chiller, for instance, if the fouling factor increases from 0.0005 to 0.0015, electrical power requirements increase by 11 percent and cooling capacity decreases by 6 percent. For absorption chillers, the same increase in fouling increases power requirements by 4 percent and decreases capacity by 20 percent (figure 3.11).[139]

One partially offsetting, positive off-stream impact of suspended sediment may result if turbidity lowers the temperature of the cooling-water supply. The cooler water could allow a smaller heat-exchange unit to be used or could increase the efficiency with which the facility operates.

As with the possible effects of turbidity on evaporation, no scientific analyses have been done on the significance or even existence of

Figure 3.11
Effects of Sediment Fouling on Cooling Efficiency

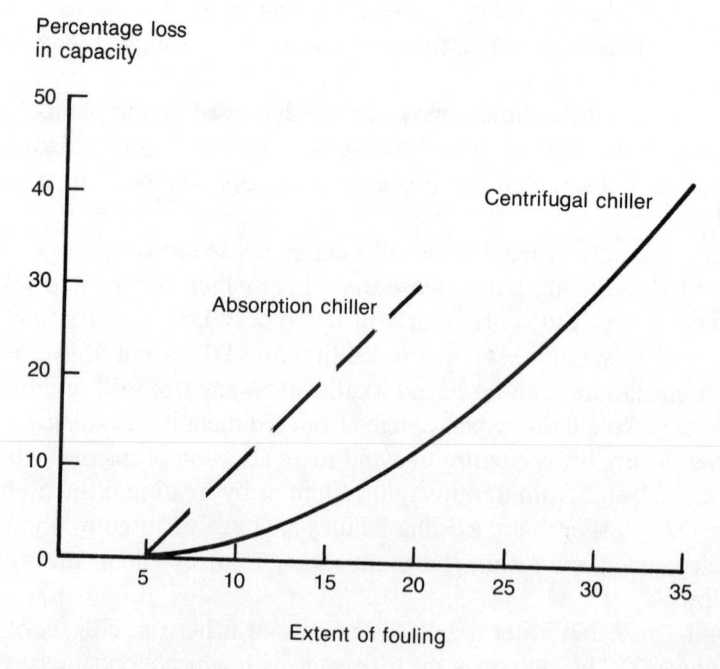

Source: Thomas A. Cady, "Removing Solid Contaminants with Side-Stream Filtration Systems," *Plant Engineering*, September 6, 1979, fig. 4, p. 120.

this possible benefit. However, an order-of-magnitude estimate of how important this factor is can be made for thermal electric power plants by using the basic laws of thermodynamics. The average efficiency of steam electric power plants currently is about 33 percent;[140] a two-degree-centigrade reduction in the temperature of the cooling water could increase the efficiency by as much as 0.4 to 0.5 percentage points.[141] In 1980, a net total of 2 trillion kilowatt hours of electricity were generated by steam electric power plants;[142] approximately 71 percent of the water withdrawn by those plants for cooling purposes came from surface freshwater sources.[143] Thus, the possible increases in efficiency could have resulted in an additional 15 to 20 billion kilowatt-hours of electricity being generated with no change in the amount of fuel used.

CONCLUSIONS

Increased sediment levels can cause a variety of downstream impacts on water quality and water use. Some of them are positive, but most are negative. Some result from the increased turbidity associated with higher amounts of suspended sediments; others are caused by sedimentation along stream channels, in lakes and reservoirs, in flooded areas, and in channels carrying water to off-stream users.

Although many of these impacts have been well documented by empirical evidence, others have not. Basic scientific and engineering principles, however, indicate that it is highly likely that the undocumented effects are occurring as well.

It is difficult to estimate the extent to which these impacts are caused by cropland erosion. The proportion of sediment that actually reaches waterways and, thus, has the potential for causing impacts is probably lower for cropland erosion than for some other major erosion sources. Erosion from stream banks certainly has a higher sediment-delivery ratio than cropland erosion has; erosion from urban areas and construction sites probably do too.*

On the other hand, because agricultural sources of eroded sediment generally are located farther upstream than are urban sources, cropland sediment probably has a higher capacity than urban erosion for causing off-site impacts. Sediment entering a river at its headwaters will

*Of course, the lower sediment-delivery ratio for cropland erosion also implies that the eroded sediment may cause significant problems on its way from a field to a waterway (for example, by blocking drainage ditches).

have opportunities to create impacts along its whole length, while sediment entering the river at its mouth can only cause impacts in the estuary. (However, if the only serious impacts are in downstream reaches, the downstream sources could be responsible for most of the problems.)

Cropland's contribution may be relatively higher for some impacts and relatively lower for others. The sediment from cropland that does reach waterways usually contains a higher percentage of fine material than the original soil did. It probably is also finer than the sediment from construction sites and stream banks. As a result, sediment from agricultural sources is more likely to remain in suspension than that from other sources. This would imply that the responsibility of agricultural sources for the impacts associated with turbidity is relatively high, while its responsibility for those impacts associated with sedimentation perhaps is somewhat lower than a straight comparison of total sediment loads would suggest.

REFERENCES

1. Kansas Department of Health and Environment, *Kansas Water Quality Management Plan as Adopted by the Kansas Legislature* (Topeka, Kans.: State of Kansas, 1979), p. IV-13.

2. J. C. Ritchie, "Sediment, Fish, and Fish Habitat," *Journal of Soil and Water Conservation*, 27 (1972):125.

3. U.S. Environmental Protection Agency, Environmental Research Laboratory, *Impacts of Sediment and Nutrients on Biota in Surface Waters of the United States*, EPA-600/3-79-105 (Athens, Ga.: U.S. Environmental Protection Agency, 1979), p. 185.

4. Vito A. Vanoni, ed., *Sedimentation Engineering,* prepared by the American Society of Civil Engineers Task Committee for the preparation of the Manual on Sedimentation of the Sedimentation Committee of the Hydraulics Division (New York: American Society of Civil Engineers, 1977), p. 348.

5. Federal Water Pollution Control Administration, *Water Quality Criteria*, report of the National Technical Advisory Committee to the Secretary of the Interior (Washington, D.C.: U.S. Government Printing Office, 1968), p. 21; and National Academy of Sciences, Environmental Studies Board, Committee on Water Quality Criteria, *Water Quality Criteria 1972*, prepared for U.S. Environmental Protection Agency (Washington, D.C.: U.S. Government Printing Office, 1973), p. 90.

6. Federal Water Pollution Control Administration, *Water Quality Criteria*, p. 46; and U.S. Environmental Protection Agency, *Quality Criteria for Water*, (Washington, D.C.: U.S. Government Printing Office, 1976), pp. 210-12.

7. Federal Water Pollution Control Administration, *Water Quality Criteria*, pp. 46-48; and Thomas R. Camp, *Water and Its Impurities* (New York: Reinhold Publishing Corp., 1963), p. 20.

8. J. O. Veatch and C. R. Humphrys, *Water and Water Use Terminology* (Kaukauna, Wisc.: Thomas Printing and Publishing Co., 1966), p. 331.

9. U.S. Environmental Protection Agency, *Quality Criteria for Water*, p. 210.

10. Ibid., p. 211; Federal Water Pollution Control Administration, *Water Quality Criteria*, pp. 46-48; and National Academy of Sciences, *Water Quality Criteria 1972*, pp. 126-29.

11. Charles E. Warren, *Biology and Water Pollution Control* (Philadelphia, Pa.: W. B. Saunders Co., 1971), p. 48.

12. Ibid., p. 48.

13. North Carolina State University, Biological and Agricultural Engineering Department, North Carolina Agricultural Extension Service, "Best Management Practices for Agricultural Nonpoint Source Control: III. Sediment," Raleigh, N.C., n.d., p. 3.

14. U.S. Environmental Protection Agency, *Impacts of Sediment and Nutrients on Biota*, p. 191.

15. R. E. Sparks, *Effects of Sediment on Aquatic Life*, (Havana, Ill.: Illinois Natural History Survey, 1977), pp. 4-5.

16. Ibid., p. 8.

17. U.S. Environmental Protection Agency, *Impacts of Sediment and Nutrients on Biota*, p. 202.

18. Ibid., p. 210.

19. R. J. Muncy et al., *Effects of Suspended Solids and Sediment on Reproduction and Early Life of Warmwater Fishes: A Review*, EPA-600/3-79-042 (Corvallis, Oreg.: U.S. Environmental Protection Agency, 1979), pp. 35-36.

20. Ibid., pp. 20-23.

21. Ritchie, "Sediment, Fish, and Fish Habitat," p. 124.

22. U.S. Environmental Protection Agency, *Impacts of Sediment and Nutrients on Biota*, p. 207.

23. Ibid.

24. Muncy et al., *Effects of Suspended Solids and Sediment*, p. 21.

25. U.S. Environmental Protection Agency, *Impacts of Sediment and Nutrients on Biota*, p. 211.

26. Sparks, *Effects of Sediment on Aquatic Life*, p. 8.

27. T. H. Langlois, "Two Processes Operating for the Reduction in Abundance or Elimination of Fish Species from Certain Types of Water Areas," *Transactions of the North American Wildlife Conference* 6 (1941):189-201; cited in Muncy et al., *Effects of Suspended Solids and Sediment*, p. 12.

28. Ibid., p. 11.

29. Michael K. Reichert, "Use of Selected Macroinvertebrates as Indicators of Sedimentation Effects on Huntington River, Utah" (M.S. thesis, Brigham Young University, 1975), p. 11.

30. Ibid.; and W. T. McClelland and M. A. Brusven, "Effects of Sedimentation on the Behavior and Distribution of Riffle Insects in a Laboratory Stream," *Aquatic Insects* 2, no. J (1980):168.

31. Ohio Environmental Protection Agency, Ohio Water Quality Management Plan, Office of the Planning Coordinator, "Scioto River Basin Agriculture Report," May 1981, revised January 1983, p. 79.

32. McClelland and Brusven, "Effects of Sedimentation on the Behavior and Distribution of Riffle Insects," p. 167.

33. Muncy et al., *Effects of Suspended Solids and Sediment*, p. 15.

34. Ibid., pp. 15, 16.

35. Ritchie, "Sediment, Fish, and Fish Habitat," p. 125.

36. Ibid.
37. Muncy et al., *Effects of Suspended Solids and Sediment*, p. 15.
38. Ibid., p. 32.
39. Ibid.
40. Ibid., p. 5.
41. M. B. Trautman, *The Fishes of Ohio* (Columbus, Ohio: Ohio State University Press, 1957); cited in Ibid., p. 38.
42. Muncy et al., *Effects of Suspended Solids and Sediment*, p. 41.
43. Ibid., p. 40.
44. N. C. Peters, "Effects on a Trout Stream of Sediment From Agricultural Practices," *Journal of Wildlife Management* 31, no. 4 (1967); Robert W. Phillips et al., "Some Effects of Gravel Mixtures on Emergence of Coho Salmon and Steelhead Trout Fry," *Transactions of the American Fish Society* 104, no. 3 (1975); F. E. Dendy, S. J. Ursic, and A. J. Bowie, "Sediment Sources and Yields from Upland Watersheds in North Mississippi" (Proceedings of the Mississippi Water Resources Conference) (Oxford, Miss.: U.S. Department of Agriculture, Sedimentation Laboratory, 1979); and see U.S. Environmental Protection Agency, *Effects of Sediment and Nutrients on Biota*, pp. 207-8 for cities.
45. J. M. Shelton and R. D. Pollock, "Siltation and Egg Survival in Incubation Channels," *Transactions of the American Fish Society* 95, no. 2 (1966):183, 184.
46. U.S. Department of Agriculture, *Science for Better Living*, The Yearbook of Agriculture 1968, House Document no. 239 (Washington, D.C.: U.S. Government Printing Office), pp. 204-8.
47. Carl L. Armour, *Effects of Deteriorated Range Streams on Trout* (Boise, Idaho: Idaho State Office, Bureau of Land Management, 1978), p. 2.
48. K. W. Dance and H. B. N. Hynes, "Some Effects of Agricultural Land Use on Stream Insect Communities," Environmental Pollution 22 (1980):26.
49. Muncy et al., *Effects of Suspended Solids and Sediment*, p. 5.
50. Robert Coats et al., *Landsliding, Channel Change, and Sediment Transport in Zayante Creek and the Lower San Lorenzo River, 1982 Water Year and Implications for Management of the Stream Resource* (Berkeley, Calif.: Center for Natural Resource Studies, 1982), p. 2.
51. Ohio Environmental Protection Agency, "Scioto River Basin Agriculture Report," p. 80.
52. North Carolina Soil and Water Conservation Commission, Division of Land Resources, and the 208 Agricultural Task Force, "Water Quality and Agriculture: A Management Plan," n.d., p. 21.
53. Muncy et al., *Effects of Suspended Solids and Sediment*, p. 26.
54. R. W. Larimore and P. W. Smith, "The Fishes of Champaign County, Illinois, as Affected by 60 Years of Stream Change," *Illinois Natural History Survey Bulletin* 28, no. 2 (1963):299-382, and P. W. Smith, "Illinois Streams: A Classification Based on Their fishes and an Analysis of Factors Responsible for Dissappearance of Native Species," *Illinois Natural History Survey Bulletin* (1971):Notes, no. 67; both cited in Ibid., p. 29.
55. Illinois Institute for Environmental Quality, "Task Force on Agriculture Nonpoint Sources of Pollution—Final Report," December 1978, pp. 47-50.
56. National Safety Council, "Accidental Deaths by Age, Sex, and Type, 1978," 1982 ed., *Accident Facts,* p. 14.

57. "Accidental Drownings by Cause and Site," Statistical Bulletin, June 1977, p. 11.

58. "Statistics of Casualties: Fiscal Year 1980," *Proceedings of the Marine Safety Council*, February 1983.

59. National Safety Council, "Total Public Deaths," *Accident Facts*, p. 74.

60. Illinois Institute for Environmental Quality, "Task Force on Agriculture Non-point Sources of Pollution," p. 41.

61. G. W. Bennett, D. H. Thompson, and S. A. Parr, "A Second Year of Fisheries Investigations at Fort Lake, 1939," Illinois Natural History Survey, Urbana, Ill., Biological notes no. 14, 1940; cited in Sparks, *Effects of Sediment on Aquatic Life*, p. 6.

62. Ibid., pp. 18, 19.

63. Ibid., p. 11.

64. Ibid.

65. John T. Daubert et al., "Economic Benefits from Instream Flow on a Colorado Mountain Stream," Colorado Water Resources Research Institute, Fort Collins, Colo., Completion Report no. 91, 1979.

66. Louis Harris and Associates, Inc., "A Survey of American Attitudes toward Water Pollution," prepared for the Natural Resources Council of America, December 1982, p. 1.

67. F. E. Dendy and W. A. Champion, *Sediment Deposition in U.S. Reservoirs: Summary of Data Reported through 1975*, prepared for U.S. Department of Agriculture, Agricultural Research Service, Misc. Pub. no. 1362 (Washington, D.C.: U.S. Department of Agriculture, 1978). This document contains a summary of data on sedimentation rates in U.S. reservoirs.

68. U.S. Department of Agriculture, *1980 Appraisal Part I: Soil, Water, and Related Resources in the United States—Status, Condition, and Trends* (Washington, D.C.: U.S. Government Printing Office, 1981), p. 221.

69. F. E. Dendy, "Sedimentation in the Nation's Reservoirs," *Journal of Soil and Water Conservation* 23, no. 4 (1968):135-7.

70. Walter B. Langbein, "Dams, Reservoirs and Withdrawals for Water Supply—Historic Trends," prepared for U.S. Department of the Interior, Geological Survey, Open-File Report no. 82-256 (Washington, D.C.: U.S. Department of the Interior, 1982), p. 2.

71. R. P. Beasley, *Erosion and Sediment Pollution Control*, (Ames, Iowa: Iowa State University Press, 1972), p. 17.

72. M. A. Churchill, "The Silt Investigations Program of the Tennessee Valley Authority" (Proceedings of the First Federal Interagency Sedimentation Conference, May 6-8, 1947), p. 53.

73. Dendy, "Sedimentation in the Nation's Reservoirs," p. 136.

74. Michael F. Walter, Tammo S. Steenhuis, and Hanneke P. DeLancey, "The Effects of Soil and Water Conservation Practices on Sediment," in Douglas A. Haith and Raymond C. Loehr, eds., *Effectiveness of Soil and Water Conservation Practices for Pollution Control*, prepared for U.S. Environmental Protection Agency, Office of Research and Development, Environmental Research Laboratory, EPA-600/3-79-106 (Washington, D.C.: U.S. Government Printing Office, 1979), p. 40.

75. Hugh Hammond Bennett, *Soil Conservation* (New York: McGraw-Hill Book

Co., 1939), p. 266.

76. Ibid., p. 267.

77. Thomas Maddock, Jr., "Reservoir Problems with Respect to Sedimentation" (Proceedings of the First Federal Interagency Sedimentation Conference), p. 13.

78. Bennett, *Soil Conservation*, p. 80.

79. Ibid., p. 271.

80. Stanley W. Trimble, "Man-Induced Soil Erosion on the Southern Piedmont, 1700-1970" (Soil Conservation Society of America, 1974), p. 119.

81. Dendy, "Sedimentation in the Nation's Reservoirs," p. 137.

82. H. G. Heinemann and D. L. Rausch, "Distribution of Reservoir Sediment—Iowa and Missouri Deep Loess Hills," in *Proceedings of the Third Federal Interagency Sedimentation Conference*, held at Denver, Colo., March 22-25, 1976 (Denver: Water Resources Council, 1976), pp. 4-144, 4-145.

83. John B. Stall, "Soil Conservation Can Reduce Reservoir Sedimentation," *Public Works*, September 1962.

84. Frank R. Schiebe, Jerry C. Ritchie, and J. Roger McHenry, "Suspended Sediment, Solar Radiation and Heat in Agricultural Reservoirs," in *Proceedings of the Third Federal Interagency Sedimentation Conference*, pp. 3-1, 3-10.

85. Maddock, "Reservoir Problems with Respect to Sedimentation," pp. 10-11.

86. Ibid., p. 11.

87. R. C. Culler et al., *Evapotranspiration Before and After Cleaning Phreatophytes, Gila River Flood Plain, Graham County, Arizona*, prepared for U.S. Department of Agriculture, Geological Survey, Professional Paper no. 655-P (Washington, D.C.: U.S. Government Printing Office, 1982).

88. Maddock, "Reservoir Problems with Respect to Sedimentation," p. 14.

89. U.S. Comptroller General, "Dredging America's Waterways and Harbors: More Information Needed on Environmental and Economic Questions," Report to Congress, CED-77-74 (Washington, D.C.: U.S. General Accounting Office, June 28, 1977), p. 1.

90. J. H. Stallings, *Soil Conservation* (Englewood Cliffs, N.J.: Prentice-Hall, 1957), p. 48.

91. U.S. Comptroller General, "Dredging America's Waterways and Harbors"; and U.S. Department of Agriculture, Soil Conservation Service, "Sediment: It's Filling Harbors, Lakes, and Roadside Ditches," prepared for U.S. Department of Agriculture, Soil Conservation Service, Agriculture Information Bulletin no. 325 (Washington, D.C.: U.S. Government Printing Office, 1967), p. 3.

92. Beasley, *Erosion and Sediment Pollution Control*, p. 19.

93. For more detailed discussions of these problems see generally U.S. Comptroller General, "Dredging America's Waterways and Harbors"; Kenneth O. Allen and Joe W. Hardy, *Impacts of Navigational Dredging on Fish and Wildlife: A Literature Review*, prepared for U.S. Department of the Interior, Fish and Wildlife Service, FWS/OBS-80/07 (Washington, D.C.: U.S. Department of the Interior, 1980); U.S. Environmental Protection Agency, Water Planning Division, "Best Management Practices Guidance—Dredged or Fill Activities," draft (Washington, D.C.: U.S. Environmental Protection Agency, n.d.); and Water Resources Council, *The Dredging Dilemma: System Problems and Management Solutions*, prepared for the New England River Basins Commission (Boston: New England River Basins Commission, 1981).

94. A. G. Taylor, "Summary of the Agriculture Task Force Water Quality Plan Recommendations," prepared by Illinois Institute for Environmental Quality, Task

Force on Agriculture Non-point Sources of Pollution, for Illinois Environmental Protection Agency, March 1978, p. 7.

95. U.S. Comptroller General, "Dredging America's Waterways and Harbors," p. 19.

96. Ibid., p. 21.

97. U.S. Army Corps of Engineers, *Deep-draft Access to the Ports of New Orleans and Baton Rouge, Louisiana—Feasibility Study*, vol. 1, *Main Report and Final Environmental Impact Statement* (New Orleans District, La.: U.S. Army Corps of Engineers, 1981), pp. 42-44.

98. Louis M. Glymph and Herbert C. Storey, "Sediment—Its Consequences and Control," in American Association for the Advancement of Science, *Agriculture and the Quality of Our Environment* (Washington, D.C.: American Association for the Advancement of Science, 1967), p. 210.

99. John J. Greiner, Jr., *Erosion and Sedimentation by Water in Texas—Average Annual Rates Estimated in 1979*, prepared for U.S. Department of Agriculture, Soil Conservation Service, Report no. 268 (Washington, D.C.: U.S. Department of Agriculture, 1982), p. 98.

100. U.S. Army Engineer Waterways Experiment Station, Environmental Laboratory, "Characterization of the Suspended Sediment Region and Bed Material Gradation of the Mississippi River Basin" (Vicksburg, Miss.: U.S. Army Corps of Engineers, Lower Mississippi Valley Division, 1981), p. E41.

101. Ibid., p. F51; and U.S. Army Corps of Engineers, *Deep-draft Access to the Ports of New Orleans and Baton Rouge, Louisiana—Feasibility Study*, p. 13.

102. "The Role of Sediment Problems in Hydroelectric Development," in *Proceedings of the Third Federal Interagency Sedimentation Conference*, pp. 41-56; and Clifford W. Randall, Thomas J. Grizzard, and Robert C. Hoehn, "Effect of Upstream Control on a Water Supply Reservoir," *Journal of the Water Pollution Control Federation* 50, no. 12 (1978):19.

103. John C. Briggs and John F. Ficke, *Quality of Rivers of the United States, 1975 Water Year—Based on the National Stream Quality Accounting Network (NASQAN)*, prepared for U.S. Department of the Interior, Geological Survey, Open Report no. 78-200 (Reston, Va.: U.S. Geological Survey, 1977), p. 41.

104. Wayne B. Solley, Edith B. Chase, and William B. Mann IV, *Estimated Use of Water in the United States in 1980*, prepared for U.S. Department of the Interior, Geological Survey, Circular no. 1001 (Reston, Va.: U.S. Geological Survey, 1983), p. 47.

105. Council on Environmental Quality, *Environmental Trends* (Washington, D.C.: U.S. Government Printing Office, 1981), p. 39.

106. E. L. Johns, "Sediment Problems in the Mohave Valley, A Case History" (Boulder City, Nev.: Bureau of Reclamation, n.d.), pp. 4-65.

107. J. J. Berger, "Flooded Village Dries Out, Packs Up, and Moves to Solar Future," Audubon 86 (May 1984):113-14 ff.

108. Tennessee Department of Public Health, Division of Water Quality Control, "Water Quality Management Plan for Agriculture in Tennessee," November 1978, p. II-7.

109. Stallings, *Soil Conservation*, p. 45.

110. Randall et al., "Effect of Upstream Control on a Water Supply Reservoir," p. 2701.

111. P. A. Kammerer, Jr., and W. G. Batten, "Sediment Deposition in a Flood-Retention Structure After Two Record Floods in Southwestern Wisconsin," *Jour-*

nal of Soil and Water Conservation 37, no. 5 (1982):302-4.

112. Fairfield Osborn, *Our Plundered Planet* (Boston: Little, Brown and Co., 1948), pp. 109-10.

113. R. D. Delaune, W. H. Patrick, Jr., and R. J. Buresh, "Sedimentation Rates Determined by Cs Dating in a Rapidly Accreting Salt Marsh," Nature 275 (1978):532.

114. Greiner, *Erosion and Sedimentation by Water in Texas*, p. 59; Trimble, *Man-Induced Soil Erosion on the Southern Piedmont, 1700-1970*; Bennett, *Soil Conservation*; Stallings, *Soil Conservation;* Rodney W. Olson and Carl L. Armour, "Economic Considerations for Improved Livestock Management Approaches for Fish and Wildlife in Riparian/Stream Areas," in *Proceedings of the Forum—Grazing and Riparian/Stream Ecosystems*, held at Denver, Colo., November 3-4, 1978 (Trout Unlimited, 1979); and E. N. Munns, "Sedimentation Problems of the Land" (Proceedings of the First Federal Interagency Sedimentation Conference).

115. Munns, "Sedimentation Problems of the Land," p. 30.

116. Trimble, *Man-Induced Soil Erosion on the Southern Piedmont, 1700-1970*, p. 51.

117. Greiner, *Erosion and Sedimentation by Water in Texas*, p. 97; and G. Y. Reddy, E. O. McLean, and G. D. Hoyt, "Effects of Soil, Cover Crop, and Nutrient Source on Amounts and Forms of Phosphorus Movement under Simulated Rainfall Conditions," *Journal of Environmental Quality* 7, no. 1 (1978).

118. Munns, "Sedimentation Problems of the Land," p. 30.

119. Beasley, *Erosion and Sediment Pollution Control*, pp. 19-20.

120. Carl B. Brown, "Perspective on Sedimentation—Purpose of Conference" (Proceedings of the First Federal Interagency Sedimentation Conference), p. 7.

121. A. G. Taylor et al., "Costs of Sediment in Illinois Roadside Ditches and Rights-of-Way," Collection of Reports on Effects and Costs of Erosion in Illinois, Illinois Environmental Protection Agency, October 1978, p. 1; and Charles Chao et al., "Nonpoint Source Pollution Studies and Control Program," prepared for Ohio Environmental Protection Agency, Office of the Planning Coordinator, June 1982.

122. Taylor et al., "Costs of Sediment in Illinois Roadside Ditches and Rights-of-Way," p. 2.

123. U.S. Department of Commerce, Bureau of the Census, *1978 Census of Agriculture*, vol. 4, *Irrigation* (Washington, D.C.: U.S. Government Printing Office, 1982), p. 266.

124. James F. Arthur and Norman W. Cederquist, "Sediment Transport Studies in the Delta-Mendota Canal and the California Aqueduct," in *Proceedings of the Third Federal Interagency Sedimentation Conference,* p. 4-88 to 4-91.

125. This calculation is relatively straightforward. The weight of sediment in 290 billion gallons of water is

$$2.9 \times 10^{11} \times 3.785 \times 10^{-4} = 1.1 \times 10^{8} \text{ kilograms per day}$$

where:

3.785 = the number of liters per gallon,

10^{-4} = the amount of sediment in kilograms per liter of water at a sediment concentration of 100 parts per million by weight.

The annual amount of energy required to lift this weight is then

$$1.1 \times 10^{8} \times 100 \times 0.3048 \times 365 \times 2.7235 \times 10^{6} \times (0.6)^{-1} = 5.5 \times 10^{6}$$

kilowatt-hours per year

where:

100 = the height of the lift in feet

0.3048 = the number of meters per foot

365 = the number of days in a year

2.7235×10^{-6} = the number of kilowatt-hours per kilogram-meter

0.6 = the average pumping efficiency.

126. Solley, Chase, and Mann, *Estimated Use of Water in the United States in 1980*, p. 10.

127. U.S. Environmental Protection Agency, *Quality Criteria for Water*, p. 210.

128. For a more detailed discussion of water treatment see Gordon Maskew Fair and John Charles Geyer, *Water Supply and Waste-water Disposal* (New York: John Wiley and Sons, 1954).

129. Richard F. McCormick et al., *Direct Filtration of Virginia Surface Waters: Feasibility and Costs* (Blacksburg, Va.: Virginia Water Resources Research Center, 1980), p. 35.

130. A. R. Castorina, "Reservoir Improvements as a Key to Source Quality Control," *Journal of the American Water Works Association* 72, no. 1 (1980):28.

131. Robert L. Morris and Lauren G. Johnson, "Agricultural Runoff as a Source of Halomethanes in Drinking Water," *Journal of the American Water Works Association* 68, no. 9 (1976):494.

132. See, generally, Castorina, "Reservoir Improvements as a Key to Source Quality Control"; Morris and Johnson, "Agricultural Runoff as a Source of Halomethanes in Drinking Water"; and U.S. Department of Agriculture, Soil Conservation Service, "Piedmont Bright Leaf Erosion Control Area—Executive Summary," n.d., p. 10.

133. National Research Council, "Chloroform, Carbon Tetrachloride, and Other Halomethanes: An Environmental Assessment (Washington, D.C.: National Academy of Sciences, 1978), p. 109.

134. John Muir Institute, *Erosion and Sediment in California Watersheds: A Study of Institutional Controls* (Napa, Calif.: John Muir Institute, 1979), p. 52.

135. Bennett, *Soil Conservation*, p. 278.

136. Mark R. Tomasch, "Cutting Hydraulics Cost with Better Filtration," *The Production Engineering Magazine* 22, no. 4 (1975):65.

137. Ibid.

138. Thomas A. Cady, "Removing Solid Contaminants with Side-Stream Filtration Systems," *Plant Engineering*, September 6, 1979, p. 117.

139. Ibid., p. 120.

140. Calculated from figures in U.S. Department of Energy, "Thermal Electric Plant Construction Costs and Annual Production Expenditures: 1980," *Monthly Energy Review*, June 1983.

141. The theoretical efficiency of a Carnot cycle is represented by the function: $(T_1 - T_2/T_1$, where T_1 is the maximum temperature and T_2 the minimum temperature in the cycle, each measure in absolute temperature. If $T_1 = 450$, $T_2 = 300$, and $T_2{}^1 = 298$, then the theoretical efficiency with the warmer cooling water is $(450 - 300)/450 = 0.3333$ and that with the cooler water is $(450 - 298)450 = 0.3378$.

142. U.S. Department of Energy, "Thermal Electric Plant Construction Costs and Annual Production Expenditures," p. 70.

143. Solley, Chase, and Mann, *Estimated Use of Water in the United States in 1980*, p. 26.

4. Impacts of Erosion-Associated Contaminants

The nutrients, pesticides, and other contaminants that are carried off agricultural lands with eroded soil also can have significant in-stream and off-stream impacts. However, unlike sediment, which is rather inert biologically, the contaminants associated with agricultural soils create impacts because of their interactions with living organisms. These effects occur through complex and sometimes indirect chemical and biological interactions rather than relatively simple physical processes, and, as a result, they often are less obviously attributable to erosion and runoff than are the impacts of sediment. The frequent indirectness, combined with the shorter history of concern about these impacts, also means that less documentation exists about the types, pervasiveness, and seriousness of the effects. However, this lack of evidence does not necessarily indicate anything about these impacts' severity or variety.

Because it often is difficult to know how the contaminant-transport process occurs—with some contaminants alternating from being adsorbed, to dissolved, and then readsorbed—little effort is made in this chapter to distinguish between the deleterious effects caused by sediment-borne pollutants and those caused by pollutants dissolved in runoff water. The magnitude and severity of a pollutant's impacts on the environment often does not depend on the particular manner in which it is carried off the land.

IN-STREAM IMPACTS

Sediment-associated contaminants can have pervasive impacts on practically every in-stream water use. Nutrients, pesticides, and other contaminants can drastically affect the quality of a water body, alter aquatic habitats, change the size and composition of a commercial or recreational angler's catch, and hinder people's recreational use of a lake or river.

Water Quality and Vegetation

The effect that a contaminant has on water quality depends on the particular type of pollutant it is and on the way in which a water body responds to it. The most severe damages occur while contaminants are dissolved or suspended in water. However, although many of the contaminants entering a water body (especially a lake or reservoir) accumulate in the bottom sediment, this deposition may not be permanent (see chapter 2). Changing physical and chemical conditions in the water body may resuspend or dissolve the contaminants, creating renewed health and environmental risks.

Eutrophication

Probably the most significant effect of eroded nutrients is the acceleration they cause in what is one of the most serious water-quality problems facing the United States—lake and reservoir eutrophication, a process manifested by excessive growth of aquatic weeds and algae. Put simply, eutrophication is the excessive nutrient enrichment of a body of water so that high rates of biological productivity are stimulated.

As plant populations "bloom" and then die and decompose, they create a rich food source for bacteria. As these bacteria decompose dead plant matter, they use up the oxygen in surrounding water. This lowers the dissolved oxygen (DO) level, especially when a cooler bottom layer of water is "locked in" beneath a warmer and heavier, oxygen-rich surface layer. Such thermal stratification usually lasts for a period of months but, in some cases, can continue for an entire year.[1]

Although eutrophication is part of the natural aging process of a water body, it is substantially accelerated when nutrients added by agricultural, residential, and industrial discharges become overabundant in the water. In its natural state, a water body usually contains inadequate amounts of one or two nutrients for the water to sustain its maximum possible plant growth. If additional amounts of those limiting nutrients are added by fertilizer runoff, plant growth is triggered and continues up to the point when another nutrient's supply becomes limiting or other constraints become operative. Phosphorus is the most common limiting nutrient, but other nutrients—nitrogen, potassium, the micronutrients—or bicarbonate can be in short supply in certain situations.

A major cause of eutrophication is runoff from agricultural fertilizer use. The nutrients that fertilizers provide to crops are often the same as the nutrients that function as limiting factors in water bodies. Particularly important in determining the effect of fertilizers on eutrophication is the extent to which the fertilizers' nutrients are readily absorbable by plants. Some nutrients are easily taken up by plants; others, for chemical or physical reasons, have much lower biological availabilities.

The extent of eutrophication is also influenced by the characteristics of the water body in which it occurs. A shallow lake receives more sunlight relative to its volume than a deep lake and, therefore, is likely to exhibit the most dramatic plant growth. In a deeper water body, excessive algae on the surface can block the sunlight, causing a reduction in the amount of beneficial bottom-weed growth and destroying the habitat of benthic organisms. The elimination of these bottom-growing plants may, in turn, further increase the levels of turbidity; these plants when living, tend to slow water currents, thereby allowing suspended sediment to settle and preventing its resuspension by waves or currents.[2]

Another factor that accelerates both unwanted surface plant and algal growth and the subsequent decomposition is increased water temperature. In addition, the larger the volume of water in a water body relative to its outflow, the greater the opportunity for nutrients to accumulate and plants to grow.

The live algae and decomposing organic matter in a eutrophic water body can cause undesirable changes in the water's color, taste, and odor. The increased photosynthesis caused by the algae can also make the water more alkaline.[3] Organisms associated with algae can add to these problems. For instance, actinomycetes, a fungus that often grows following a blue-green algal bloom, produces a chemical compound called geosmin, which causes a particularly disagreeable, intense, earthy smell. (This odor can remain after water is withdrawn and processed for drinking.)

If the decomposition of excess vegetation uses up all the available oxygen in a lake or reservoir's bottom, it can also increase the concentrations of iron and manganese compounds in the water.[4]* The

---

*When all the readily available oxygen is depleted, decomposing algal compounds satisfy their continuing need for oxygen by chemically reducing manganese oxide hydroxides and iron oxide hydroxides in the bottom sediments. The resulting iron and manganese are released into the water.

release of these heavy metals, like the lowered oxygen level, can severely disturb the populations of benthos, hampering their ability to control harmful coliform bacteria and toxic compounds.

Advanced stages of eutrophication are accompanied by prolific growth of emergent plants and the encroachment of edge species that can decrease a lake or reservoir's water-storage capacity. Under natural conditions, transition from an open lake to a marsh and, eventually, a dry bed takes thousands of years. However, if no efforts are taken to control weed growth, increased nutrient loadings can accelerate the process significantly. The increased plant growth would be expected to increase the rate of water loss through transpiration.* In addition, to the extent that the plants occupy space that would otherwise be filled by water, they may cause the water level to rise, increasing the surface area and, as a result, the amount of evaporation.

On balance, the effects of eutrophication are definitely negative. However, especially in the early stages, it can sometimes perform some positive functions. For example, overall dissolved oxygen levels may increase if the total photosynthetic and respiratory rates (processes that give off oxygen) for a lake's aquatic vegetation are higher than the oxygen-consumption rates for its decomposing dead vegetation. And, if turbidity levels are low, the eutrophication can stimulate, rather than supress, bottom-weed growth. In such cases, the extra vegetation may, in addition to stabilizing bottom sediments: (*a*) trap small insoluble pollutants and absorb inorganic particles such as nitrates, phosphates, and metals in their stems and roots, (*b*) reduce the number of pathogenic bacteria in the water possibly by producing chemicals toxic to the bacteria, and (*c*) provide shelter for various beneficial bacteria, fish, and insects that aid in removing contaminants from the water.[5] When the eutrophication primarily stimulates the growth of algae and surface plants, the increased opacity of the water is likely, similar to sediment-caused turbidity, to lead to reduced sunlight absorption and, therefore, lower surface-water temperatures (see chapter 3).

Increased Salinity

Dissolved salts, transported from cropland by irrigation return flows and runoff, also can threaten water supplies. Throughout the Southwest, salt cedar, an exotic salt-tolerant plant that consumes large

---

*Although both of these effects would be the expected results of eutrophication, no previous analysis that was reviewed in preparing this study identified them as serious problems.

amounts of water, is rapidly encroaching on many streams and lakes.[6] The salt cedar appears to thrive in areas (such as the Canadian River in Texas) where a water body's salinity interferes with the propagation of more salt-sensitive, less thirsty plants that would normally compete with it.

Salts, however, can reduce the toxicity of some other compounds—a potentially beneficial effect. The toxicity of many heavy metals and pesticides in water depends on variables such as temperature, hardness, acidity, presence of organic compounds and of other metals, and DO levels.[7] For instance, dissolved solids such as carbonates, magnesium, and, in particular, calcium can decrease heavy metal toxicity.

Wildlife

Nutrients, pesticides, and other agricultural contaminants all reduce the amount and diversity of aquatic wildlife in a body of water. However, the mechanisms by which they create these problems differ substantially from one another.

Nutrients

Some forms of nutrients in runoff harm aquatic organisms directly. Ammonium, for instance, can be acutely toxic to fish at concentrations of two parts per million (ppm) in highly alkaline waters. Concentrations higher than this have been measured in runoff from agricultural fields.[9]

Most nutrient-related damages to aquatic communities, however, are a result of eutrophication and are, therefore, indirect. Increased nutrient loadings in a water body change its species composition and dramatically increase the quantity of microscopic plants it contains. For instance, in Lake Erie, phytoplankton domination shifted from diatoms, minute brown or clear single-celled algae that are an important food source for aquatic invertebrates and that do not make water green or smelly, to less desirable blue-green and green algae and to nuisance growths of *Cladophora*, a filamentous alga that attaches to shallow lake bottoms.[10]

Sometimes these algal blooms can make the fish within a water body toxic. Shellfish, which siphon through their systems whatever water surrounds them, can be particularly affected by blooms of toxic algae such as the "red tides" of dinoflagellates that excessive nutrient runoff from coastal lands has occasionally caused in marine fisheries. These blooms have occurred in areas as dispersed as the New Jersey and Florida coasts, the Adriatic Sea, and Japan's Seto Inland Sea.[11]

A recent U.S. Environmental Protection Agency (EPA) research report on the Chesapeake Bay documents the loss of submerged vegetation from nutrient enrichment as one of the most serious problems facing the bay.[12] The report correlates the buildup of nutrients in northern estuarine reaches of the bay with the disappearance of grasses and submerged vegetation from the same areas. Light sufficient for submerged plant growth penetrates to a depth of only one meter in most of the northern Chesapeake. Consequently, these plants have been limited to shallow waters where other stresses, such as wave action, waterfowl grazing, and extreme variations in water temperature hinder their growth.[13]

This loss of vegetation can have serious repercussions throughout a food chain. Submerged leafy plants and grasses are both a sheltering habitat and an important food source for many animals. The plants support the growth of plankton and insect larvae. Juvenile fish feed on the larvae and hide from their predators. Adult fish, the predators, find the vegetation a rich foraging area. Likewise, estuarine grasses are nurseries for the larvae of crustaceans and shellfish. In the southern Chesapeake, where vegetation has not yet disappeared, eelgrass beds are primary nurseries for juvenile blue crabs.[14]

The detritus (organic trash) from decaying aquatic plants is also an important direct food source for many zooplankton, insect larvae, amphipods, shrimp, and fish. Thus, the loss of aquatic vegetation affects not only these animals but also their predators, such as sea bass, pipefish, pigfish, and white perch.[15]

Moreover, loss of submerged aquatic plants reduces species diversity for several reasons. Vegetated areas attract and sustain a much greater species abundance and diversity than do unvegetated areas. In addition, the algal growths caused by eutrophication often result in valuable sport and commercial fish such as perch and trout being replaced by "rough" fish such as carp, goldfish, and gizzard shad, which are generally considered less desirable. The turbidity caused by the suspended phytoplankton, like that caused by suspended sediment, inhibits the more highly valued fish in their search for prey. And, once carp are established as a dominant species, they perpetuate turbid conditions by stirring up bottom sediments as they root about for food.[16]

The decomposition of algae and other vegetation can combine with other biochemical oxygen demand (BOD) loadings to reduce oxygen levels in water, thereby harming animals ranging from microscopic zooplankton to large game fish. In Lake Erie, for instance, eutrophication-caused oxygen depletion in the cooler, deeper water is held responsible for destroying the benthos and making much of the area uninhabitable for many fish and their food species (figure 4.1). This, along with overfishing and siltation of their spawning grounds, caused the extinction of lake trout in Erie.[17] Eutrophication's effects have also been chronicled in North Carolina, where severe algal blooms and fish kills in the Chowan Estuary are blamed for costly losses to the fishing and tourist industries. Soil erosion and other nonpoint sources contribute 85 percent of the Chowan River's nitrogen load.[18]

The problem of winterkill is also more common in eutrophic lakes. When ice and snow cover a lake surface, not enough sunlight penetrates the water to stimulate the growth of oxygen-producing algae. Nevertheless, the bacterial decomposition of organic material left over from warmer weather continues, gradually consuming the water's entire oxygen load and killing the fish.[19]

However, under certain circumstances, eutrophication can increase fish yields, particularly during the high productivity of the early eutrophication stage.[20] For instance, eutrophication resulting from decay of submerged terrestrial vegetation during the first few years of a reservoir's life usually provides excellent fishing.[21]

Figure 4.1
Suspected Causes of Decreased Fish Populations in Lake Erie

	Species						
Cause	Lake trout	Sturgeon	Northern pike	Lake herring	Whitefish	Sauger	Blue pike
Related to erosion							
Low dissolved oxygen[1]	X	X		X	X		X
Siltation	X				X	X	
Other causes							
Over-exploitation	X	X		X	X	X	X
Tributary damming		X	X			X	
Marsh drainage		X	X				
Temperature flucuations				X	X		X
Hybridization						X	X

[1] Reduced levels of dissolved oxygen during summer months caused in part by accelerated eutrophication.

Source: Justine Welch, *The Impact of Inorganic Phospates in the Enviroment,* prepared for U.S. Environmental Protection Agency, Office of Toxic Substances (Washington, D.C.: U.S. Environmental Protection Agency, 1978), table 6, p. 58.

The in-stream impacts of nutrients are not only limited to aquatic wildlife. For much of the year, estuaries along the East Coast, including the Chesapeake, are important feeding grounds for nesting and migratory waterfowl. Submerged aquatic vegetation is a major food item for these birds.[22] Its disappearance will at least cause the birds to go elsewhere and may result in reduced bird populations. There are fewer waterfowl in the Chesapeake region, but they may be compensating for the Chesapeake's decline in vegetation by wintering elsewhere along the Atlantic Flyway migration route.[23] Similarly, predatory birds such as herons, ospreys, bald eagles, and kingfishers will be adversely affected by reductions in the fish population in areas suffering from excessive nutrient enrichment.

Pesticides

Extensive research has documented the effects of many pesticides on a wide variety of living organisms. These can be loosely grouped into two categories: acute effects, which produce an immediate and violent response such as death, and chronic effects, which persist at low levels or may not appear for a long period of time. Pests for which a pesticide was developed usually suffer its acute effects, while nontarget birds, fish, and people are more likely to suffer chronic effects, often from indirect exposure.

In most surface waters, pesticide concentrations are too diluted to produce dramatic effects; however, when storms or accidental spills dump large quantities into a stream or lake, massive fish kills can result. From 1961 through 1975, agricultural pesticides were responsible for almost one-fourth of all reported fish kills for which the cause of the problem was known (figure 4.2).[24] In 1976, individual incidents of pesticide runoff from agricultural lands killed a reported 109,000 fish in an Alabama fish hatchery, 20,000 fish in Marine World Lake in California, 17,500 fish in Idaho's Rock Creek, and almost 30,000 fish in Illinois's Middle Branch. In all, 79 incidents that year were reported to have killed almost 400,000 fish, with most of the incidents resulting in kills of 500 to 3,000 fish each. In several other cases, pesticides were reported as the cause of a fish kill, but there was no estimate of the number of dead fish.[25] In Louisiana, concentrations of the herbicide paraquat that are likely to occur in runoff have been found to kill juvenile crayfish, an important local food source.[26]

Chronic effects are more common, and, even when the concentrations are not high enough to produce chronic effects, some pesticides can bioaccumulate in animal bodies to unsafe levels for those who consume them. Pesticides may interfere with an animal's normal functions such as respiration, reproduction, and locomotion, threatening its long-term survival. Even very low concentrations of certain pesticides are suspected of interfering with the basic metabolic processes—enzyme activity, oxygen consumption, and growth rates—of fish, birds, and

Figure 4.2
Number of Fish Killed Annually by Source of Pollution, 1961–1980 (in Thousands)

Source	Average, 1961–75	1976	1977	1978	1979	1980
Agricultural						
Pesticides	1,500	390	560	450	680	330
Fertilizers	48	110	69	72	400	450
Manure-silage drainage	520	140	76	270	420	1,300
Subtotal	2,000	640	710	800	1500	2,100
Other sources						
Industrial	7,000	590	700	2,400	1,500	1,100
Municipal	14,000	10,000	1,300	2,500	450	6,300
Transportation	450	580	640	250	630	870
Miscellaneous	3,900	640	210	3,100	510	18,000
Subtotal	25,000	12,000	2,800	8,200	3,100	26,000
Total—Known sources	27,000	13,000	3,500	9,000	4,600	28,000
Total—Unknown sources	5,900	870	13,000	66,000	3,500	1,600
Total	**33,000**	**14,000**	**17,000**	**75,000**	**8,100**	**30,000**

Columns do not add up precisely, because all numbers are rounded to two significant figures.

Source: U.S. Environmental Protection Agency, *Fish Kills Caused by Pollution, Fifteen-Year Summary 1961–1975*, EPA-440/4-78-011 (Washington, D.C.: U.S. Government Printing Office, 1979), table 3, p. 8.; U.S. Environmental Protection Agency, *Fish Kills Caused by Pollution in 1976*, EPA-440/4-79-024 (Washington, D.C.: U.S. Government Printing Office, 1979), table 2, p. 5; and U.S. Environmental Protection Agency, "Fish Kills by Source of Pollution, 1977–1980" (unpublished).

mammals.[27] In fish, some pesticides are known to cause delayed sexual maturation, reduced reproductive efficiency, testes regression, and reductions in the hatching of eggs.[28]

The degree of bioaccumulation that is possible depends both on the pesticide and on the kind of organism. Pesticides (particularly organochlorines) with low water solubilities and high fat solubilities tend to concentrate in the fatty tissues of animals that ingest them.[29] Such pesticides also tend to adsorb readily onto sediments and suspended organic materials, making them a particularly serious problem for filter-feeding species such as shellfish. Because they are unable to avoid environmental contaminants, shellfish can ingest and bioaccumulate pesticides up to 70,000 times the concentration in the surrounding water.[30] Similarly, fish that eat detritus along the bottom sediments of lakes and rivers are likely to have higher than average concentrations of pesticides in their tissues. And these pesticides may persist in the sediment long after their use on land has been stopped.

Some pesticides also can be passed along to higher organisms in the food chain, making them particularly harmful. A well-known example is DDT (dichloro-diphenyl-trichloro-ethane), which "biomagnified"—that is, it became increasingly concentrated as it moved up the aquatic food chain, until it accumulated at very high levels in fish-eating eagles and ospreys and caused them to lay thin-shelled eggs that could not hatch.

Moreover, pesticides can affect aquatic species indirectly by reducing their food supply or destroying their natural habitat. Herbicides washed off by storms may kill submerged aquatic vegetation found in lakes, ponds, and estuaries. The herbicide atrazine, for instance, has been found to at least temporarily reduce photosynthesis in many aquatic plants in the Chesapeake Bay.[31] Herbicides such as cyanatryn, used to control algae and plants like Eurasian millfoil and water hyacinth in drainage ditches, have been found to kill off snails, daphnia, and other invertebrates that were important food sources for fish.[32]

Other Contaminants

Increased concentrations of dissolved solids can also affect an aquatic habitat. Different aquatic organisms have different tolerances for these salt ions. Whitefish and pikeperch can survive concentrations of up to 15,000 milligrams per liter, while sticklebacks can survive concentrations of up to 20,000 milligrams per liter.[33] Increased salinity, of course, favors the more salt-tolerant species, stimulating changes in the types and populations of different species throughout an aquatic food chain. And the effects are not limited to aquatic wildlife. High summer salt concentrations in California's Suisun Marsh between 1956 and 1960 poisoned young ducklings, resulting in retarded growth or death.[34]

Fluctuations in salinity, however, may cause more problems than absolute salt concentrations.[35] Fish can sometimes adapt to constant high concentrations of dissolved solids. Although EPA considers lakes with greater than 15,000 milligrams per liter of dissolved solids unsuitable for freshwater fishes,[36] both fish and waterfowl have adapted to the much higher but constant salinity level in the Great Salt Lake in Utah.

Other harmful impacts that salinity can have on wildlife habitat include the destruction of habitat and food source plants and, as wetlands along waterways become saline and dry up, the destruction of riparian habitat.[37] Finally, the different salinity levels (gradients) in estuaries are used by many organisms as a protective haven from predators; small fish, for example, go into less-saline waters that their predators cannot tolerate. These gradients can be destroyed by highly saline runoff waters entering the bays[38] or, conversely, by greater than normal freshwater runoff that dilutes the estuarine waters.

Metals associated with erosion also can affect an aquatic environment. Even though these metals exist at low concentrations in the water, they accumulate in bottom sediments and can pose a risk to bottom-feeding organisms and their predators. Such animals as filter-feeding shellfish and sediment-dwelling worms ingest the metals and concentrate them in their tissues, causing reduced fertility, reproductive decline, and shortened life spans.[39] As various species fall victim to the metals in a water body, the balance of the benthos is upset, and the ecosystem can become unstable as species further up the food chain see a decrease in their food supply. Because these materials bioaccumulate and biomagnify, their concentrations may be much higher in the animals than in the sediment. However, many of the documented instances of metals having serious effects on wildlife are associated with sources of pollution other than cropland erosion.[40]

Commercial Fishing

The adverse effects of agricultural contaminants on aquatic wildlife can seriously affect commercial fishing. In Lake Erie, for instance, a marked decline in the commercial fishing industry has accompanied its serious eutrophication problems. Since 1962, the number of commercial fishers employed by the industry declined from an average of 963 to 392 and the gross tonnage of fishing vessels dropped from 1,388 to 488.[41] As a result, the number of processing and wholesale manufacturing plants for fish products in Ohio and Pennsylvania dropped from 73 to 33 between the 1950s and the 1970s.[42] Although the business losses attributable to eutrophication are impossible to quantify, agricultural contaminants clearly have had a significant effect. The destruction of feeding and spawning habitat in Lake Erie initially made commercial fishing easier by forcing schools of fish to concentrate in smaller remaining suitable areas. After the high-value fish were depleted, however, fishing pressure had to shift to lower value species. For instance, as whitefish, blue pike, and lake herring gradually disappeared, yellow perch catches increased.[43]

In some places where pesticide contamination has occurred, commercial fishing has been prohibited. Commercial exploitation of some Great Lakes fish was banned in the mid-1970s because of unsafe levels of pesticides in their tissues.[44] The highly prized Coho salmon of Lake Michigan were restricted from sale in interstate commerce in 1968, due to unacceptably high DDT levels. Lake trout, lake herring, and some chubs were likewise banned for commercial consumption.[45]

Pesticides, even when present at safe levels in fish tissues, may affect the flavor of certain fish. In one study, professional taste testers ate rainbow trout that had been exposed to varying levels of the herbicides acrolein and 2,4-D. The testers found that concentrations of 0.1 milligrams per liter of acrolein and 0.5 milligrams per liter of 2,4-D in the water rendered the fish "unpalatable."[46] The unpleasant taste in the fish persisted for a week after the pesticide had been added to the water, even after the chemicals themselves were no longer at detectable concentrations. Similarly, the fish-farming industry of Manitoba's prairie lakes was damaged by blooms of *Oscillatoria*, a blue-green alga common to eutrophic waters. The algae produced geosmin, giving fish in the lakes an objectionable, muddy flavor.[47]

Shellfish, because they are stationary filter feeders, unable to avoid polluted zones, may be particularly affected by these problems. In 1980, die-offs of juvenile shrimp following toxaphene, methyl parathion, and endosulfan applications in coastal counties led South Carolina's Shrimpers Association to express its concern over the "possibility of great damage" to the area's commercial fishing industry.[48] And the tendency of shellfish to take up toxic algae that bloom in summer months may be the source of the folk wisdom that prohibits eating shellfish during months without "Rs" (May through August).[49] Contamination by fecal coliform in runoff can also be a problem, especially if grazed pastureland is adjacent to shellfish waters.[50]

Fortunately, pesticide concentrations measured in fish have declined significantly since the more persistent pesticides such as DDT and dieldrin were banned.[51] In most fish, the concentrations are now within limits EPA considers safe for human consumption.

Dissolved solids and other contaminants affecting aquatic habitats also can hurt commercial fisheries. Estuaries may be most vulnerable to pollution damage, being both highly productive, richly diverse waters for commercial fishing and recipients of much of the pollution coming from the land, with relatively little turnover of water to flush out contaminants. Many commercial species of fish depend on the different levels of salinity an estuary provides to trigger their various life stages. For instance, according to a National Academy of Sciences report, shad "depend upon fresh water areas at the head of estuaries for spawning and for survival as eggs and larvae, open estuaries for the nutrition of juveniles, and large open coastal regions for growth and maturation." Salinity changes at any of these areas could "break the necessary patterns and reduce the fishery."[52]

Finally, heavy metals and other toxic substances that settle out of water can harm commercial fishing. Many commercial species are benthos (for example, shrimp, crabs, and oysters) that may be most affected by these contaminants. Other species that do not live on the bottom may depend on those animals that do for their food supply.

Recreation

The many impacts of the associated contaminants on aquatic wildlife obviously decrease the suitability of the waters for recreational as well as commercial fishing. And these impacts usually reinforce those caused by sediment—for instance, by accelerating the replacement of game fish by hardier but less desirable species and by making the waters more turbid and thus obscuring the angler's lures.[53]

Some highly eutrophic, turbid lakes produce large populations of stunted panfish that have little fishing value. Since their predators are unable to see them, large numbers of these fish survive to adulthood but, competing with each other for food and space, do not grow as large as they might otherwise.[54]

In addition, pesticides in water can affect recreational fishing in much the same way that they affect commercial fishing. In the Great Lakes, for instance, people fishing for sport were warned in 1979 that they should not eat more than one meal of fish per week because of unsafe pesticide concentrations; in some cases, the possession of certain kinds of fish was prohibited.[55] And, although methods have been devised for trimming and cooking fish in ways that reduce pesticide levels,[56] those methods also reduce tastiness.

Such contaminants can also affect other recreational activities. Algae are probably responsible for some of the turbidity-related swimming and boating accidents described in chapter 3. Swimmers, divers, water-skiers, and scuba divers generally regard dense stands of aquatic vegetation as a nuisance. In extreme cases, such aquatic plants contribute to drownings, as when they obstruct a diver's view of submerged hazards or become entangled in scuba gear.[57]

Serious health problems can result from swimming in eutrophic waters if certain species of algae have become abundant. Toxic effects on humans attributable to algae include gastrointestinal, respiratory, and dermatological disorders.[58] In the highly eutrophic lakes of western Canada, cases of severe diarrhea, vomiting, and other discomforts have been reported for people who swam in or swallowed water heavily infested with *Microcystis* and *Anabaena* algae.[59] *Oscillitoriae*-infested waters have been known to produce respiratory disorders in swimmers, and *Anabaena* blooms have also given swimmers dermatological problems.[60] Some people experience asthma, dermatitis, and other allergic responses when exposed to algae-infested waters. Only a couple of cases of algal poisoning have been reported for people, although livestock have been killed by drinking water heavily infested with blue-green algae.[61]

Even when a body of water is only tangential rather than central to a recreational activity, as with hiking or picnicking, people prefer a clear, clean lake or stream to a murky, green one, as mentioned in chapter 3. People's enjoyment of a lake or stream cannot help but be reduced if it is clogged with floating mats of algae or it emits the sulfurous stench of rotting vegetation that follows an algal bloom.

Shorelines and shallow water can become choked with large, weedy aquatic plants that interfere with boating, fishing, and swimming. The extent of interference varies with such factors as the growth form of the plants, how densely they colonize, and how much of the water surface they cover.[62] Strands of floating macrophytes can wind around a motorboat propeller, jamming it, and can hopelessly entangle fishing lines and lures. For instance, eutrophic conditions producing excessive growth of duck weed, hydrilla, and hyacinth have interfered with pleasure boating in Texas's Lake Livingston Reservoir.[63] In addition, floating mats of algae or free-floating macrophytes can wash up onto recreational beaches, where their subsequent drying and decay produces objectionable odors and attracts insects.[64]

Pesticides also can affect people's enjoyment of such recreational activities as bird-watching, snorkeling, wildlife photography, camping, and hiking because of the harm they do to aquatic plants, fish, and other wildlife.

Similarly, high salinity levels in water bodies can decrease the suitability of a lake or stream for recreational purposes. The increased evaporation rates from saline waters may decrease the amount of water available for boating, fishing, or swimming. Moreover, the salinity, by affecting the species and populations of aquatic wildlife, may diminish the attractiveness of recreational fishing. And, in water bodies used for swimming, even slight changes in salinity can cause skin and eye irritations. For instance, changes in the water's pH, which can be caused by dissolved solids, may result in severe eye irritation.[65]

For the most part, because heavy metals are tied up in bottom sediments, they are of less concern to recreationists, although these contaminants can affect sport fishing. Further, heavy metal concentrations can favor the development of hardy strains of pest insects such as the midge, making them resistant to normal pest-control techniques.[66]

Other In-Stream Impacts

Contaminants carried in water can affect several other in-stream uses as well. Excess vegetation resulting from eutrophication can clog water intakes and distribution pipes for hydroelectric facilities.[67] And, depending on the circumstances, increased salinity may also cause these facilities to corrode; chlorides, for instance, may take the form of hypochlorous acid and act as an oxidizing agent.[68]*

Other in-stream damages by polluted water can include accelerated depreciation of bridges, wharfs, and piers; blockage of navigation routes by sediments; and increased repair costs for ships and other floating vessels.[70] Most of these damages probably are not due to agricultural contaminants, however.[71]

OFF-STREAM IMPACTS

When water is withdrawn from a stream, lake, or reservoir, the contaminants carried along with suspended sediment interfere with off-stream uses in several ways. Communities taking water for their drinking supply, factories and power plants that need water for production processes or to cool machinery, and farmers diverting streams for irriga-

*Other pollutants, however, can reduce corrosiveness. Water with low levels of dissolved oxygen or with moderate or high levels of alkalinity is less corrosive than "pure" water.[69] In addition, under certain conditions, dissolved solids *decrease* water's corrosive effects, as when the addition of basic salt anions neutralizes corrosive acidic ions already present in the water.

tion all bear added costs if the water is contaminated.

Lakes and reservoirs can be particularly affected as suitable sources of water for off-stream uses, since many contaminants tend to accumulate in such water bodies. Yet from 30 to 40 percent of the U.S. population depends on lakes and reservoirs for its water supply.[72] One study, produced in 1978 for EPA, reported that nonpoint-source pollution in South Dakota lowered the quality of drinking water for more than 12 percent of the state's population.[73]

Drinking-Water Supplies

Agricultural pollutants can cause problems for public water supplies at practically every stage in the process, from collection through treatment and delivery to water users. Many water-treatment plants lack the capacity to remove contaminants such as nitrates and pesticides, and even those that can remove normal levels of contaminants may not be able to deal with the higher concentrations that can appear after a major spring rainstorm. A large number probably lack even the capacity to test adequately for the presence of such contaminants in the water supply.

Nutrients

Nitrogen's and phosphorus's impacts on drinking-water supplies can be both direct, as when a nutrient itself is in a toxic form or combines to form dangerous compounds, and indirect, as when it stimulates algal growth, which in turn hampers water purification.

Although most of the nitrogen and phosphorus compounds commonly found in water supplies are not dangerous to human health, nitrites are. They can cause methemoglobinemia, a temporary blood disorder in infants that causes oxygen deprivation and sometimes suffocation by inhibiting hemoglobin's ability to carry oxygen throughout the bloodstream.[74] The nitrite concentration in water usually is low, but intestinal bacteria can convert nitrates, found in much higher concentrations, into nitrites. However, most of the 2,000 recorded cases of methemoglobinemia in Europe and the United States have resulted from contaminated rural wells; only one case has been attributed to a public water supply.[75] To prevent the disease, EPA recommends a limit of 10 milligrams per liter of nitrates as a drinking-water standard.[76]

In addition, another nitrogen compound, ammonium, is found in surface water at high enough concentrations to cause concern for human health. Ammonium in agricultural runoff has been measured to exceed EPA's drinking-water standard of 0.5 ppm.[77] Ammonium also corrodes copper and copper-alloy pipes and fixtures, stimulates algal growth in water-distribution systems, and reduces the effectiveness of chlorine disinfectants.[78]

Most drinking-water supply problems, however, result from eutrophication, as was indicated by a 1966 survey of state sanitary engineers. That survey identified eutrophication as affecting 56 percent of the total surface-water supply used for domestic purposes in this country.[79] Eutrophication can increase the cost and difficulty of supplying adequate water at every step of the collection/treatment/distribution process.

Probably the most serious problems occur at the water treatment plant. Algae can clog or disrupt water filters used in removing contaminants. At West Germany's Wahnbach Reservoir, for instance, algae filaments frequently break through the sand filter and into the water-supply system.[80] And a November 1972 through January 1973 bloom of the tiny alga *Melosira islandica* so clogged the filters that, after four hours of operation, the capacity of the filtration system was reduced by 30 percent.[81] Algal clogging is also common in the Cleveland, Ohio, water-supply system.[82] An August 1974 clogging by *Oedegonium*, a filamentous alga, was so severe that the city's Division Avenue water-filtration plant shut down for three days, leaving many of its 400,000 users without water.[83] Filter clogging used to be such a problem for Belleview, Ontario, at the eastern end of Lake Ontario, that the city equipped its treatment plant with a microscreen to strain large algae out of the intakes.[84]

To avoid such problems, most plants attempt to remove algae by adding chemicals that cause the algae and suspended sediments to flocculate and settle out of the water before it reaches the filters. However, the algae, water-soluble organic compounds that the algae produce, and other organic compounds carried in the water can seriously interfere with the flocculation process. Some of the compounds are strongly acidic and will enrich the negative surface charges of particles that need to be removed, rather than neutralize their charges, as is required for flocculation to be successful. Other organic substances form insoluble compounds with the chemicals added as part of the flocculation process, necessitating the use of much greater quantities of such chemicals. Tests on water taken from Wahnbach Reservoir during one algal bloom showed that over 10 times as much aluminum sulfate was required to reverse the flocculant's negative charge as would have been required with algae-free water.[85]

The presence of phosphorus can further complicate flocculation problems: pyrophosphate and triphosphate salts have been found to interfere with coagulation and flocculation, again requiring the use of greater amounts of coagulant.[86]* Such large chemical demands both increase treatment costs substantially and generate large amounts of sludge that require disposal.

Even elaborate, well-operated processes of flocculation and filtration are not 100 percent effective in removing algae. Extremely efficient treatment plants may remove 99.9 percent of diatoms; more typical, however, is the 90 percent elimination rate achieved for the blue-green alga *Oscillatoria rubescens*. The lower efficiency rate allows significant amounts of the algae or their decomposition products to pass into the distribution systems and storage tanks. There, the deposited organic matter forms a rich medium for the growth of bacteria and other aquatic species.[87]

Inadequate treatment often results in smelly, oily-tasting water. Taste and odor problems were by far the most serious alga-associated treatment problem recognized by state sanitary engineers in the 1966 study.[88] Many species of algae themselves have disagreeable tastes and odors; more often, however, it is the products of algal metabolism or decomposition that cause problems described as "aromatic, spicy, garlic, pigpen, fishy, grassy, musty, earthy, and septic."[89] Taste and odor problems for Lake Michigan water users are attributed to the algae *Dinobryon* and *Uroglenopsis*; similar problems with Lake Erie water are blamed on the blue-green alga *Anabaena*.[90] In northern Virginia's Occoquan Reservoir, such problems coincide with high growths of *Anabaena, Microcystis,* and *Aphanizomenon*.[91]

Activated carbon filters can be added to a water-treatment system in an attempt to remove substances causing taste and odor problems. But even this was only partially effective in dealing with a massive bloom and subsequent decomposition of the diatom *Melosira italica* in the Wahnbach system. The fishy, oily smell and taste in the drinking water persisted for over four weeks after the bloom was over because the offensive compounds remained trapped in the filter.[92]

Even the basic disinfection of water is hampered by the presence of suspended and dissolved compounds. Organic colloids can form protective coatings around microorganisms, preventing disinfectants from penetrating and killing the germs.[93] Even with longer disinfection periods and the use of larger quantities of chemicals, the disinfection process may be incomplete. If, as is usually the case, chlorine is used as the disinfectant, the high chemical concentrations can in

---

*Such complex phosphorus compounds are more likely to be associated with point-source pollution (that is, from detergents) than with fertilizer runoff.

themselves affect the taste and odor of water and, of far greater health concern, create toxic compounds such as trihalomethanes and chlorite ions.[94] Trihalomethanes such as chloroform, estimated to comprise about 20 percent of the organochlorines in drinking water, are a suspected carcinogen for humans. The other 80 percent of the organochlorines have not yet been clearly identified but may include even more hazardous compounds.[95]

With eutrophic water supplies, the disinfection stage is not the only point where extra chlorine is added to the water. It may also be added to intake water to prevent the growth of organisms in intake pipes; during treatment to control the growth of bacteria, algae, and zooplankton and to break down ammonia and other substances; and in the distribution system to control growth there. In addition, extra chlorine may be added to precipitate high levels of substances such as iron and magnesium that result from eutrophic conditions. The Crown filtration plant in Cleveland, for instance, has on occasion added three to five times the normal amount of chlorine to deal with high concentrations of these metals.[96] All of this additional chlorine can produce trihalomethanes and other organochlorinated compounds.

Pesticides

Pesticide concentrations associated with agricultural runoff and erosion rarely are high enough to produce acute effects such as convulsions or death. However, too little is understood about the levels capable of generating the known chronic effects of pesticides—such as cancer, miscarriages, and mutations—to permit unconcern when even low concentrations reach water supplies.[97]

People with occupational exposure to specific pesticides are subject to a wide variety of diseases, including hypertension, hyperlipoprotenemia, soft tissue sarcomas, histocytic lymphomas, neurologic and behavioral abnormalities, sterility, neurotoxicity, and cancer.[98] Some pesticides or their by-products have been found to cause genetic mutation, and normal metabolic processes can create mutagens out of substances that appear to be safe in their original form (figure 4.3). (Corn, for instance, takes up atrazine from the soil and converts it into a mutagenic compound.[99]) And, in some cases, a contaminant or by-product of the production process is responsible for mutagenesis, such as dioxin associated with the pesticide 2,4-D and nitrosamines in trifluralin.

But such effects have been documented only in cases of relatively high exposure, such as may occur in occupational situations. Occur-

Figure 4.3
Apparent Mutagenicity of Selected Herbicides

Compound	Relative mutagenic effect	Plant activitated[1]
Sodium azide	large	yes
Atrazine	some	yes
Simazine	some	no
Cyanazine	some	yes
Diquat	not tested	
Paraquat	not tested	
2, 4-D	none observed	
—Dioxin contaminant by-product	extremely large	no
Dicamba	none observed	
Trifluralin	none observed	
—Nitrosoamine by-product	large	no
Linuron	none observed	
Alachlor	some	yes
Propachlor	some	yes

[1] Indicates whether plant metabolism can convert original material into mutagenic substance.

Source: U.S. Environmental Protection Agency, "Chesapeake Bay Program Technical Studies: A Synthesis" (Washington, D.C.: U.S. Environmental Protection Agency, 1982), table 4, p. 537.

rences of high pesticide concentrations in water supplies appear to be fairly infrequent and localized. By the time water reaches a customer's tap, pesticide concentrations are seldom, if ever, at levels that are thought to produce health effects, and even irrigation return flow usually does not contain pesticide concentrations higher than the federal drinking-water standards.[100] However, these standards may not define guaranteed "safe" levels because so much remains unknown about the long-term health effects of even very small concentrations. Likewise, little is known about the synergistic effects among different pesticides and between pesticides and other substances.

Most water-treatment plants are not equipped to treat for pesticides. For instance, when pesticide concentrations in the raw water supply and in the treated drinking water were monitored for the city of Tiffin, Ohio, concentrations were found to be very similar, indicating that the city's purification treatment removed virtually none of the seven pesticides commonly used throughout the watershed.[101] Although the measured concentrations were not acutely toxic, some (including atrazine) were found at levels sufficiently high to inhibit plant growth in the rivers.

Many facilities even lack the diagnostic equipment necessary to identify pesticides in the supply. The high costs of specialized equipment and specially trained personnel required for pesticide detection prevents most smaller communities from routine monitoring, leaving the job to large cities, federal and state agencies, and private laboratories.[102]

Finally, the trend for currently used agricultural pesticides to be more water soluble than were the organochlorine and organophosphate pesticides used in the 1960s and early 1970s may increase the problems surrounding their removal by water-treatment facilities.* Whereas chlorinated hydrocarbons like DDT could, to some extent, be immobilized by sediment particles and removed by filtering and sedimentation basins, removing soluble pesticides is much more difficult and expensive.

Other Contaminants

Many dissolved solids—calcium, sodium, bicarbonate magnesium, chloride ions—also can negatively affect domestic water supplies. High concentrations impart an objectionable mineral taste to drinking water and, for some people, have a laxative effect.[104] In addition, excess sodium can create hypertension and adversely affect cardiac patients and pregnant women.[105]

The recommended drinking-water limit for total dissolved solids (TDS) is 500 ppm.[106] By comparison, the combination of heavy salt loadings and high evaporation rates has caused the dissolved solid concentrations in many shallow North Dakota lakes to reach 240,000 milligrams per liter, over seven times the concentration of sea water.[107] In Pueblo County, Colorado, concentrated salinity from the evaporation and transpiration of irrigation return flows has frequently increased TDS concentrations in drinking-water sources up to four times the recommended water-quality standards for domestic use.[108]

An increase in dissolved solids can pose other problems for water users as these solids corrode or form deposits in distribution pipes, on various utensils, and in household fixtures. To avoid these problems, households must attempt to avoid the problems by installing water softeners. Otherwise, they either must attempt to remove the deposits or replace equipment when it becomes too corroded or the deposits on it become too great.[109]

---

*As noted in chapter 2, although these compounds are generally less persistent than their predecessors, the persistence of some pesticides is much greater in water than on a field. Aldicarb persists up to 20 years in water, compared to a half life in the soil of a few weeks.[103]

Industrial Water Use

Water-borne agricultural contaminants also present problems for industries. Industrial use (including steam electric power plants and miscellaneous uses) accounts for almost 58 percent of all water withdrawals in the United States, though only 8 percent of water consumption.[110] According to one estimate, "more than 20 percent of the gross water intake by American industries is subject to controls for biological growth, removal of suspended or dissolved solids, corrosion control, and other abatement processes."[111]

In 1968, about 28 percent of the industrial water withdrawn in the United States (excluding withdrawals for thermoelectric power) was used for processing. This water, depending on the industry and the use to which it is put, may need to be of drinking quality or better. If industries such as food canning and soft drink production must process and purify their own water supplies, they face the same problems that agricultural nutrients and pesticides present to municipal water-treatment plants.[112]

Approximately 90 percent of industrial withdrawals are used for cooling where high quality water is not needed. Nevertheless, nutrients and other contaminants can cause excessive algal growth in cooling towers and in recirculation ponds. Power plants along Lake Brawning in Texas, for instance, report reduced condenser efficiency as a result of excessive algal growth on the condenser walls. However, such algal growth usually can be controlled relatively easily by "shock chlorination"—brief treatments of high concentrations of chlorine.[113]

As with drinking-water systems, industries and power plants are plagued by clogging and fouling of filters at intake points. Weeds, algae, and rough fish are primarily responsible. As eutrophication of a supply source progresses, routine maintenance and cleaning costs can be expected to increase correspondingly.[114] During certain seasons, power plants on Lake Ontario require daily removal of *Cladophora* algae and other trash (up to 5.6 cubic meters) to ensure adequate water supply. The lakeshore town of Pickering, Ontario, eventually had to install expensive heat exchangers in its power plant to control *Cladophora*, which had persistently clogged water lines, occasionally to the point of forcing the plant to shut down.[115]

High concentrations of dissolved solids also can pose serious problems for industrial water users. These salts are incompatible with many physical and chemical processes for which water is needed and often result in scaling and corrosion of machinery. Further, treatment processes to remove the salts are not always available. Removing these contaminants is further complicated because the water also contains organic particles and suspended solids associated with phytoplankton growth. Such substances often have to be removed before the ion-exchange process of demineralization can take place.

Irrigation

The impacts of biological and chemical contaminants on irrigated agriculture vary depending on such factors as crop type, soil type, climate, topography, and farmland slope. In general, contaminants adversely affect crops by suppressing vegetative growth, reducing fruit development, or decreasing their quality as a marketable product.[116]

Nutrients in irrigation water probably have predominantly beneficial effects. For instance, every acre-foot of water containing 1.0 milligrams per liter of nitrogen and 0.5 milligrams per liter of phosphorus contributes the equivalent of 5.8 pounds of urea and 7 pounds of superphosphate fertilizer.[117] However, the indirect effects of these nutrients sometimes can be negative. Nuisance aquatic plants from nutrient-rich reservoirs and canals can be carried in irrigation water and can then spread into fields and overtake them. For instance, *Hydrilla*, "the Godzilla of aquatic plants,"[118] has been found to outcompete rice crops in both Florida and California.[119]

The impact of other contaminants on irrigated crops is predominantly negative. For instance, croplands can be infested by plant pathogens carried off an upstream farm—an effect that has been documented at least for the reuse of irrigation return flows in eastern Washington and California,[120] where nematodes (a wormlike pest that lives in the soil) affecting root growth and development have spread from field to field.

Pesticides also may cause problems. They can change the chemical composition, appearance, texture, and flavor of marketable crops, lowering their value.[121] Residual herbicides carried in water may diminish a plant's growth. For instance, the herbicide endothall, used to control algal blooms in reservoirs, has been found to injure corn, soybeans, and sugar beets.[122]

But the major problem for irrigated cropland is probably caused by salts, particularly in the arid West. High levels of sodium can cause leaf burn and render crops unmarketable.[123] Chlorides are harmful to citrus, stone fruits, and almonds.[124] High levels of soluble sulfate cause browning in some lettuce varieties,[125] rendering the crops less suitable for marketing.

Increased salt concentrations in irrigation water may decrease crop yields by 20 percent.[126] In the Colorado River system alone, annual damages attributed to salinity ranged from $75 million to $104 million in 1980 and are predicted to increase to $165 million in the year 2000.[127]

High soil salinity creates a "physiological drought" condition. As salts accumulate in the soil, damages increase and large amounts of water may be needed to flush the salts out of the root zone. For instance, up to 20 percent more water is required to irrigate crops in saline areas of the Great Plains.[128] In some soils, salts react with the soil constituents to decrease permeability, making irrigation water less accessible to plants.[129] And, if a surface crust forms, seed germination and seedling emergence is deterred.[130]

The particular effects of salinity and other contaminants often depend on the chemistry of the contaminant and the soil, as well as the physical and chemical characteristics of the irrigation water. These may interact in a synergistic or antagonistic mannner. For instance, soils that become too acidic lose their natural capacity to absorb contaminants.[131] And the salts in the irrigation water can cause some soils to flocculate. The resulting increased permeability causes the nutrients to be carried rapidly out of the root zone.[132]

It has been estimated that, unless current irrigation practices are changed, the ability of many lands to sustain crops will drop severely, with up to 100,000 acres lost to production in the 1980s.[133] Moreover, the current mechanisms for offsetting these damages are expensive. The most common response is for a farmer to use more of some input—more water to offset the effects of salinity, or more pesticides to offset the binding of pesticides by organic materials in the irrigation water.[134] In some cases farmers plant crops that are resistant to the contaminants. A particularly expensive response has been the construction of a $216 million nuclear-powered desalinization facility in Yuma, Arizona, to allow the United States to satisfy its Colorado Treaty obligations with Mexico.[135]

Other Off-Stream Impacts

Other off-stream uses of water can be affected by the contaminants carried in water. Some pollutants—especially dissolved solids—will corrode or leave encrustations on pipelines and other water-conveyance facilities. Nutrients can stimulate weed growth in drainage ditches (though other contaminants may reduce such growths). It is also possible that the damages and health risks associated with flooding may be increased by these contaminants' presence in the floodwaters. However, in the research for this report, no studies documenting the existence of such effects, much less their magnitude, were found.

CONCLUSIONS

T'.is chapter has identified several negative and some positive impacts that can be caused by the various contaminants associated with cropland soil erosion. All the caveats listed in the conclusion of chapter 3 apply even more emphatically here. In many cases, the only information available indicates no more than that a particular type of impact can be caused by contaminants associated with soil erosion, not that soil erosion itself is causing such impacts. In other cases, even less information is available—perhaps only enough to support a hypothesis that such impacts may exist.

This paucity of evidence results primarily from three factors. First, relatively slight attention has been given in past research to problems caused by contaminants associated with soil erosion. Second, it is difficult to determine which of the many different sources of these contaminants—including industries, municipal treatment plants, and urban stormwater runoff, as well as soil erosion—are responsible for any particular impact that is observed. Similarly, identifying the correct causative agent or agents—pesticides, low dissolved oxygen concentrations, overfishing, "natural" population dynamics, or any of a number of other possibilities—is seldom easy. Finally, most of these contaminants produce the impacts indirectly. Nutrients, for instance, stimulate algal growth that, in turn, harms fish or increases drinking-water treatment costs. The impacts tend to result from chemical or biological interactions that are more difficult to observe than are sediment's impacts. Siltation of a spawning bed is more easily observed, for instance, than is a fish population decline caused by pesticides' elimination of its invertebrate food source. Whether or not the specific causes of a particular impact are known, assigning each factor its share of the responsibility for the total problem is close to impossible.

Even if it were possible to estimate the magnitude of these problems and the contribution that different types of contaminants make to them, it still would be very difficult to determine how much of each contaminant comes from soil erosion. Only rough estimates exist of how much of the different types of contaminants come from agricultural land. Moreover, eliminating all erosion would not keep some unknown proportion of this unknown amount from entering the waterways through groundwater leaching and surface runoff.

The lack of information on all these factors is so overwhelming that it may never be possible to document the precise contribution of cropland erosion to the impacts described in this chapter. Still, this does not mean that the impacts are unimportant or that the contribution of cropland erosion is trivial. As this chapter has indicated, many impacts may be very serious indeed.

REFERENCES

1. National Academy of Sciences, Environmental Studies Board, Committee on Water Quality Criteria, *Water Quality Criteria 1972*, prepared for U.S. Environmental Protection Agency, EPA-R3-73-033 (Washington, D.C.: U.S. Government Printing Office, 1973), p. 19.

2. U.S. Environmental Protection Agency, "Chesapeake Bay Program Technical Studies: A Synthesis" (Washington, D.C.: U.S. Environmental Protection Agency, 1982), p. 410.

3. Werner Stumm and James J. Morgan, *Aquatic Chemistry: An Introduction Emphasizing Chemical Equilibria in Natural Waters*, 2d ed. (New York: John Wiley and Sons, 1981), p. 17; and W. D. Billings, *Plants and the Ecosystem* (Belmont, Calif.: Wadsworth Publishing Co., 1964), p. 50.

4. National Academy of Sciences, *Water Quality Criteria 1972*.

5. William A. Hanson and Frans Bigelow, *Lake Management Case Study: Westlake Village, California,* Technical Bulletin no. 73 (Washington, D.C.: Urban Land Institute, 1977), p. 76.

6. Panhandle Regional Planning Commission, "Plan Summary Report for the Canadian Basin Water Quality Management Plan," prepared for Texas Department of Water Resources, August 1978, revised June 1981, p. II-D-11.

7. Black & Veatch, Consulting Engineers, "Toxic Substance Pollutant Sampling Program," prepared for Lower Rio Grande Valley Development Council 208 Water Quality Program, Dallas, Texas, June 1981, p. VI-2.

8. U.S. Environmental Protection Agency, *Quality Criteria for Water* (Washington, D.C.: U.S. Government Printing Office, 1976), p. 76.

9. J. L. Baker, "Agricultural Areas as Nonpoint Sources of Pollution"; cited in Michael R. Overcash and James M. Davidson, eds., *Environmental Impacts of Nonpoint Source Pollution* (Ann Arbor, Mich.: Ann Arbor Science Publishers, 1980), pp. 298-99.

10. Justine Welch, *The Impact of Inorganic Phosphates in the Environment*, prepared for U.S. Environmental Protection Agency, Office of Toxic Substances, EPA-560/1-78-003 (Washington, D.C.: U.S. Environmental Protection Agency, 1978), p. 9.

11. G. F. Lee and R. A. Jones, "Effect of Eutrophication on Fisheries," prepared for American Fisheries Society, Engineering Resources Center (Fort Collins, Co.: Colorado State University, 1981), pp. 12-13.

12. U.S. Environmental Protection Agency, "Chesapeake Bay Program Technical Studies," p. 410.

13. Ibid., p. 487.

14. Ibid., p. 489.

15. Ibid., p. 488.

16. Lee and Jones, "Effect of Eutrophication on Fisheries," pp. 7-8.

17. Welch, *The Impact of Inorganic Phosphates in the Environment,* p. 59.

18. North Carolina Soil and Water Conservation Commission, Division of Land Resources, and the 208 Agricultural Task Force, "Water Quality and Agriculture: A Management Plan," n.d., p. 21.

19. Lee and Jones, "Effect of Eutrophication on Fisheries," pp. 10-11.

20. Ibid., p. 1.

21. U.S. Environmental Protection Agency, *Restoration of Lakes and Inland Waters* (Proceedings of the International Symposium on Inland Waters and Lake Restoration, Portland, Maine, September 8-12, 1980), EPA 440/5-81-010 (Washington, D.C.: U.S. Environmental Protection Agency, 1980), p. 358.

22. U.S. Environmental Protection Agency, "Chesapeake Bay Program Technical Studies," p. 488.

23. Ibid., p. 449.

24. U.S. Environmental Protection Agency, Office of Water Planning and Standards, *Fish Kills Caused by Pollution, Fifteen Year Summary 1961-1975*, EPA-440/4-78-011 (Washington, D.C.: U.S. Government Printing Office, 1979), p. 8.

25. U.S. Environmental Protection Agency, Office of Water Planning and Standards, *Fish Kills Caused by Pollution in 1976*, EPA-440/4-79-024 (Washington, D.C.: U.S. Government Printing Office, 1979), pp. 5, 55-63.

26. Tat-Sing Leung, Syed M. Naqvi, and Nusrat Z. Naqvi, "Paraquat Toxicity to Louisiana Crayfish (Procambarus clarkii)," *Bulletin of Environmental Contamination and Toxicology* 25, no. 465 (1980):465-69.

27. H. Singh and T. P. Singh, "Short-Term Effect of Two Pesticides on the Survival, Ovarian ^{32}P Uptake and Gonadotophic Potency in a Freshwater Catfish, *Heteropneustes Fossils (Bloch)*," *Journal of Endocrinology* 85 (1980):193; and Erik G. Ellgaard et al., "An Analysis of the Swimming Behavior of Fish Exposed to the Insect Growth Regulators, Methoprene and Diflubenzuron," *Mosquito News* 39, no. 2 (1979):311-19.

28. Singh and Singh, "Short-Term Effect of Two Pesticides on the Survival, Ovarian ^{32}P Uptake and Gonadotophic Potency in a Freshwater Catfish," p. 194.

29. R. L. Metcalf and J. R. Sanborn, "Pesticides and Environmental Quality in Illinois," *Illinois Natural History Survey Bulletin* 31, no. 9 (1975):402.

30. P. A. Butler, "Pesticides in the Marine Environment," *Journal of Applied Ecology* 3 (1966); cited in National Academy of Sciences, *Water Quality Criteria 1972*, p. 37.

31. U.S. Environmental Protection Agency, "Chesapeake Bay Program Technical Studies," p. 557.

32. H. R. A. Scorgie and A. S. Cooke, "Effects of the Triazine Herbicide Cynatryn on Aquatic Animals," *Bulletin of Environmental Contamination and Toxicology* 22 (1979):137, 141.

33. U.S. Environmental Protection Agency, *Quality Criteria for Water*, p. 207.

34. National Academy of Sciences *Water Quality Criteria 1972*, p. 195.

35. Ibid.

36. U.S. Environmental Protection Agency. *Quality Criteria for Water*, p. 207.

37. U.S. Department of the Interior, Bureau of Reclamation and U.S. Department of Agriculture, Soil Conservation Service, "Final Environmental Statement, Colorado River Water Quality Improvement Program," vol. 1, FES 77-15, May 19, 1977, p. I-38.

38. U.S. Environmental Protection Agency, *Quality Criteria for Water*, p. 207.

39. Ibid. See discussions of individual metals.

40. See Edward D. Goldberg et al., "The Mussel Watch," *Environmental Conservation* 5, no. 2 (1978):101; Overcash and Davidson, *Environmental Impacts of Nonpoint Source Pollution*, pp. 298-99.

41. Welch, *The Impact of Inorganic Phosphates in the Environment*, p. 65.

42. Ibid., p. 66.

43. Ibid., pp. 61-63.

44. National Research Council, Committee on Impacts of Emerging Agricultural Trends on Fish and Wildlife Habitat, *Impacts of Emerging Agricultural Trends on Fish and Wildlife Habitat* (Washington, D.C.: National Academy Press, 1982), p. 212.

45. National Academy of Sciences, *Water Quality Criteria 1972*, p. 184.

46. L. C. Folmar, "Effects of Short-Term Field Application of Acrolein and 2,4-D (DMA) on Flavor of the Flesh of Rainbow Trout," *Bulletin of Environmental Contamination Toxicology* 24, no. 217 (1980):217-24.

47. J. L. Tabachek and M. Yurkowski, "Isolation and Identification of Bluegreen Algae Producing Muddy Odor Metabolites, Geosmin, and 2-Methylisoborneol in Saline Lakes in Manitoba," 1976; cited in Welch, *The Impact of Inorganic Phosphates in the Environment*, p. 48.

48. Letter from D. M. Gilliken, president of South Carolina Shrimpers Association to James A. Timmerman, Jr., executive director of the State of South Carolina Wildlife and Marine Resources Department, April 11, 1980.

49. U.S. Environmental Protection Agency, *Restoration of Lakes and Inland Waters*, p. 357.

50. R. W. Skaggs et al., *Effect of Agricultural Land Development on Drainage Waters in the North Carolina Tidewater Region* (Raleigh, N.C.: North Carolina Water Resources Research Institute, 1980).

51. Council on Environmental Quality, *Environmental Trends* (Washington, D.C.: U.S. Government Printing Office, 1981), p. 264; and Christopher J. Schmitt et al., *National Pesticide Monitoring Program: Organochlorine Residues in Freshwater Fish, 1976-1979*, Resource Publication no. 152 (Washington, D.C.: U.S. Fish and Wildlife Service, 1983), p. 23.

52. National Academy of Sciences, *Water Quality Criteria 1972*, p. 221.

53. U.S. Environmental Protection Agency, *Restoration of Lakes and Inland Waters*, p. 167.

54. Lee and Jones, "Effect of Eutrophication on Fisheries," p. 8.

55. J. C. Skea et al., "Reducing Levels of Mirex, Aroclor 1254, and DDE by Trimming and Cooking Lake Ontario Brown Trout (*Salmo trutta Linnaeus*) and Smallmouth Bass (*Micropterus dolomieui Lacepede*)," *Journal of the Great Lakes Resource* 5 (1979):153.

56. Skea et al., "Reducing Levels of Mirex, Aroclor 1254, and DDE," p. 154.

57. National Academy of Sciences, *Water Quality Criteria 1972*, p. 26.

58. U.S. Environmental Protection Agency, *Restoration of Lakes and Inland Waters*, p. 167.

59. Ibid., p. 167.

60. Ibid., p. 167.

61. Welch, *The Impact of Inorganic Phosphates in the Environment*, p. 17.

62. National Academy of Sciences, *Water Quality Criteria 1972*, p. 25.

63. Trinity River Authority, "Plan Summary Report for the Trinity Basin Water Quality Management Plan," prepared for the Texas Department of Water Resources, July 1978, revised June 1981, p. II-D-36.

64. National Academy of Sciences, *Water Quality Criteria 1972*, p. 26.

65. Ibid., p. 33.

66. JACA Corporation, *An Assessment of Economic Benefits of 28 Projects in the Section 314 Clean Lakes Program*, prepared for U.S. Environmental Protection Agency, Office of Water Planning and Standards (Fort Washington, Pa.: JACA Corporation, 1980), p. 2.

67. National Academy of Sciences, *Water Quality Criteria 1972*, p. 376.

68. Thomas R. Camp, *Water and Its Impurities* (New York: Reinhold Publishing Corp., 1963), p. 189.

69. Ibid., pp. 159-160.

70. Fred H. Abel, Dennis P. Tihansky, and Richard G. Walsh, "National Benefits of Water Pollution Control," prepared for U.S. Environmental Protection Agency, Washington Environmental Research Center, Office of Research and Development (Washington, D.C.: U.S. Environmental Protection Agency, n.d.), pp. 36, 38.

71. Ibid., p. 38.

72. JACA Corporation, *An Assessment of Economic Benefits of 28 Projects in the Section 314 Clean Lakes Program*, p. 2.

73. South Dakota Department of Environmental Protection, Office of Water Quality, "The South Dakota Statewide 208 Water Quality Management Plan," November 1978, p. 25.

74. G. Stanford, C. B. England, and A. W. Taylor, "Fertilizer Use and Water Quality," prepared for U.S. Department of Agriculture, Agricultural Research Service, no. ARS 41-168, October 1970, p. 2.

75. National Academy of Sciences, *Water Quality Criteria 1972*, p. 73.

76. Ibid., p. 73.

77. Baker, "Agricultural Areas as Nonpoint Sources of Pollution," p. 298.

78. National Academy of Sciences, *Water Quality Criteria 1972*, p. 55.

79. American Water Works Association, "Nutrient-Associated Problems in Water Quality and Treatment," *Journal of the American Water Works Association* 58 (1966):1348.

80. U.S. Environmental Protection Agency, *Restoration of Lakes and Inland Waters,* p. 360.

81. Ibid., p. 361.

82. Ibid., p. 167.

83. Welch, *The Impact of Inorganic Phosphates in the Environment*, p. 46.

84. U.S. Environmental Protection Agency, *Restoration of Lakes and Inland Waters*, p. 170.

85. Ibid., p. 362.

86. American Water Works Association, "Nutrient-Associated Problems in Water Quality and Treatment," p. 1338.

87. U.S. Environmental Protection Agency, *Restoration of Lakes and Inland Waters*, p. 360.

88. American Water Works Association, "Nutrient-Associated Problems in Water Quality and Treatment," p. 1349.

89. Ibid., p. 1343.

90. Welch, *The Impact of Inorganic Phosphates in the Environment*, p. 47.

91. Ibid., p. 48.

92. U.S. Environmental Protection Agency, *Restoration of Lakes and Inland Waters,* p. 361.

93. North Carolina State University, Biological and Agricultural Engineering Department, North Carolina Agricultural Extension Service, "Best Management Practices for Agricultural Nonpoint Source Control: III. Sediment," Raleigh, N.C., n.d., p.2.

94. U.S. Environmental Protection Agency, *Restoration of Lakes and Inland Waters*, p. 362.

95. Ibid., pp. 374-375.

96. Welch, *The Impact of Inorganic Phosphates in the Environment*, p. 46.

97. John A. Miranowski, "Agricultural Impacts on Environmental Quality"; cited in Ted L. Napier et al., eds., *Water Resources Research: Problems and Potentials for Agriculture and Rural Communities* (Ankeny, Iowa: Soil Conservation Society of America, 1983), p. 129.

98. John E. Davies and Virgil H. Freed, eds., "An Agromedical Approach to Pesticide Management, Some Health and Environmental Considerations," prepared in cooperation with the Agency for International Development Consortium for International Crop Protection, University of Miami, School of Medicine, undated draft, pp. 5-6.

99. M. J. Plewa and J. M. Gentile, "Mutagenicity of Atrazine: A maize-microbe bioassay," *Mutation Research* 38 (1976):287-92); cited in Maureen K. Hinkle, "Problems with Conservation Tillage," *Journal of Soil and Water Conservation* 38, no. 3 (1983):204.

100. National Academy of Sciences, *Water Quality Criteria 1972*, pp. 346-348.

101. Charles Chao et al., "Nonpoint Source Pollution Studies and Control Program," prepared for Ohio Environmental Protection Agency, Office of the Planning Coordinator, June 1982, p. 16.

102. National Academy of Sciences, *Water Quality Criteria 1972*, p. 76.

103. "Water Contaminated by Pesticide Spurs Growing Concern in Florida," *New York Times*, March 28, 1983.

104. National Academy of Sciences, *Water Quality Criteria 1972*, p. 90.

105. North Dakota State Department of Health, Division of Water Supply and Pollution Control, "North Dakota Statewide 208 Water Quality Management Plan," Water Quality Report, March 1979, p. 12.

106. Gordon Maskew Fair and John Charles Geyer, *Water Supply and Waste-Water Disposal* (New York: John Wiley and Sons, 1954), p. 570.

107. North Dakota State Department of Health, "North Dakota Statewide 208 Water Quality Management Plan," p. 104.

108. Pueblo Regional Planning Commission, "Executive Summary for Pueblo County: Agricultural Water Quality Assessment of Lower Fountain Creek," prepared for Pueblo Area Council of Governments, March 1982, p. 13.

109. U.S. Department of the Interior and U.S. Department of Agriculture, "Final Environmental Statement, Colorado River Water Quality Improvement Program," p. 206.

110. Wayne B. Solley, Edith B. Chase, and William B. Mann IV, *Estimated Use of Water in the United States in 1980*, U.S. Geological Survey Circular no. 1001 (Reston, Va.: U.S. Department of the Interior, Geological Survey, 1983), p. 33.

111. Abel, Tihansky, and Walsh, "National Benefits of Water Pollution Control," p. 43.

112. National Academy of Sciences, *Water Quality Criteria 1972*, pp. 369, 392.

113. Welch, *The Impact of Inorganic Phosphates in the Environment*, pp. 48-49.

114. Ibid., p. 49.

115. Ibid., p. 49.

116. National Academy of Sciences, *Water Quality Criteria 1972*, p. 324.

117. Ibid., p. 348.

118. Denis Collins, "Solution to 'Monster' Hydrilla May Present Another Problem," *Washington Post*, July 15, 1984, p. F6.

119. JACA Corporation, *An Assessment of Economic Benefits of 28 Projects in the Section 314 Clean Lakes Program*, p. B-38.

120. National Academy of Sciences, *Water Quality Criteria 1972*, p. 348.
121. Ibid.
122. Ibid., p. 347.
123. Ibid., p. 329.
124. Ibid.
125. Ibid., p. 327.
126. Joel Grossman and Sandra Marquardt, "Salting of the Earth," *Environmental Action*, June 1983, p. 30.
127. U.S. Department of the Interior and U.S. Department of Agriculture, "Final Environmental Statement, Colorado River Water Quality Improvement Program," p. I-21.
128. M. L. Quinn, "Strategies for Reducing Pollutants from Irrigated Lands in the Great Plains," Nebraska Water Resources Center, University of Nebraska-Lincoln, Robert S. Kerr Environmental Research Laboratory, July 1, 1982, p. 45.
129. National Academy of Sciences, *Water Quality Criteria 1972*, p. 324.
130. Ibid.
131. Ibid., p. 339.
132. Ibid., p. 324.
133. Grossman and Marquardt, "Salting of the Earth," p. 30.
134. Quinn, "Strategies for Reducing Pollutants from Irrigated Lands in the Great Plains," p. 111.
135. Grossman and Marquardt, "Salting of the Earth," p. 30.

5. Economic Costs: Some Estimates

Only a few efforts to estimate the financial magnitude of the off-farm impacts of soil erosion have ever been made. The first Federal Interagency Sedimentation Conference, held in 1947, placed the total cost of seven major types of damage at an estimated $175 million (approximately $630 million in 1980 dollars).[1]* A 1979 review estimated a cost of $1.12 per acre just for dredging sediment from roadside drainage ditches, waterways, harbors, and reservoirs in the United States—equivalent to approximately $1.2 billion nationwide.[2]

Some studies have looked at the economics of avoiding these damages or correcting them once they have occurred. A 1974 study of a 3,000-acre watershed in Illinois concluded that adopting soil conservation measures would cost less than the damages caused by sedimentation in drainage ditches and reservoirs and by flooding.[3] The cost of those damages went as high as $20.10 per cropped acre for some sections of farmland ($31 per cropped acre in 1980 dollars), depending on cropping patterns and the particular control measures adopted.

No similar cost-benefit comparisons are made in this chapter. However, a step in that direction is taken by attempting to provide a national estimate of the monetary value of the off-site damages caused by sediment and other erosion-related contaminants.

The science (or, perhaps more appropriately, the art) of estimating the economic costs of environmental damages still is relatively crude. When there are reliable measures of the physical damage, and when the items affected have a clear value in the market, damage estimates are fairly straightforward. In many cases, however, both the extent of

*All cost estimates in this chapter are both given as they were originally presented and converted to 1980 dollars. The implicit gross-national-product deflator is used for converting to 1980 dollars unless some other price index is indicated in a footnote.

physical damage and the value that should be placed on the damaged item are unclear. This is certainly the case with damages caused by water pollutants. Typically, the damages are difficult to measure and depend on the specific water body affected and the particular uses that are made of it. Many of the uses involve activities that normally take place without market exchanges—that is, people rarely purchase the right to fish or to look at pleasant landscape. Determining the economic value of such activities is very difficult.

For these reasons, the cost estimates presented in this chapter are very uncertain. For each damage category, the extent of uncertainty is expressed by indicating a range of costs. For instance, the estimated range of costs associated with removing sediment from drainage ditches is given as $90 to $185 million per year. Several ranges are even broader—an order of magnitude or more.* After the range is estimated for a category, a "single-value" estimate is chosen. This value lies within the range and represents a best guess of the approximate extent of the damages. The range and the single-value estimate relate to damage caused by all sediment and associated contaminants, regardless of their source.

The final step is to estimate what share of these damages can be attributed to cropland erosion. Several estimates of pollutant loadings indicate that cropland erosion contributes about one-third of the sediment and a comparable amount of the nutrients entering streams.[4] Therefore, unless a contrary assumption is indicated, cropland is assumed to account for one-third of the damages. This is admittedly a very crude assumption. There are many reasons why cropland's share should be higher in some cases and lower in others. In most cases, however, the available data do not justify choosing any other particular percentage, and thus the one-third assumption is retained.

In some cases, no cost estimate can be made on the basis of presently available information. In addition, some resources affected by soil erosion do not participate in the economic market, which serves to indicate the value of resources to people. As a result, it is not possible to assign them any value. In addition, it is generally harder to quantify the impacts of erosion-related contaminants such as nutrients and pesticides than of sediment. It should not be assumed, however, that those impacts for which no estimates are made are less important than those for which a number is provided. Just because the cost of an impact cannot be estimated does not make the impact any less important.

---

*An order of magnitude means that one value is ten times greater than another value.

The cost estimates contained here should be greater than the actual benefits that would result from controlling erosion and runoff. One reason, associated primarily with sediment-related damages, is the automatic effort of alluvial streams to maintain equilibrium. If the sediment carried into such a stream were substantially reduced, the stream would begin to erode its own banks to bring the combination of stream slope, sediment load, and water volume back into balance. During the period when this adjustment process was occurring—in most cases, for several years—the increased sediment resulting from stream-bank erosion would partially compensate for the reduced sediment entering the stream from cropland erosion. As a result, the downstream damages would continue to occur even though the cropland erosion would have been reduced.

Second, erosion-control measures could not stop all erosion. Though the damages should be substantially reduced, some of them would continue to occur even after a stream had regained an equilibrium.

Finally, the costs of the damages may have already been paid before any erosion-control measures were adopted. As a result, little real savings would result from a reduction in damages. For instance, if a new lakeside marina had been built to replace one that had "silted out," abandoning the new facility and returning to the original site would likely generate additional net costs, not net benefits.

Given all these problems, prudence might suggest abandoning any effort to estimate monetary costs. In this case, prudence was rejected. Although many of the costs estimated in this chapter may be little more than crude guesses, they do serve to provide some indication of how serious the damages are compared to other national priorities.

IN-STREAM IMPACTS

The preceeding two chapters identified five primary in-stream effects of soil erosion—impacts on aquatic ecosystems, recreation, water storage, navigation, and commercial fishing. All of the just-described estimation problems are encountered when attempts are made to determine the approximate costs of these impacts.

Biological Impacts

Although the aquatic organisms that are affected by the off-farm impacts of soil erosion generally do not participate in the economic market, their ecosystems do provide recreational opportunities and food for humans. Aquatic organisms can be very important in other ways as well. They can contribute to scientific and medical knowledge. They serve as a genetic library that may be valuable in future efforts to breed for disease resistance and other beneficial characteristics. They may help eliminate human or natural wastes occurring in water or may recycle valuable substances, making them available again for human use. And, on another level, wildlife can help monitor the health of the general environment: fish kills and other significant changes in an aquatic ecosystem can serve as a warning of possible serious human health effects associated with water pollution.

These are only some of the values attributable to natural systems and their individual members.[5] Most biologists agree that, to assure the maintenance of these benefits, diverse, complex ecosystems must be preserved. The more diverse an ecosystem is, the more stable it likely will be and the better it will be able to absorb stresses without collapsing. For this reason, each member of an ecological community is important in maintaining stability, even though that member may have no apparent direct benefit to humans.

Most estimates of the economic importance of types of wildlife assign values on the basis of the benefits they currently provide humans for recreation, commercial fishing, and industrial raw materials. These types of estimates are discussed in other sections of this chapter. Clearly, however, such approaches miss many of the less measurable values described above. Yet, recognizing these potential benefits, the United States sometimes has undertaken substantial expenditures to protect endangered species—including spending over $16 million to preserve the Mississippi sandhill crane, $20 million to preserve the California condor, and $7 million to preserve the whooping crane.[6]

A few efforts to assign values to members of an ecosystem have been made. For instance, the Florida Department of Pollution Control, in its attempts to estimate damages caused by water pollution, assigns specific dollar values to members of different aquatic species (figure 5.1).[7] But the bases for the Florida values are not clear. They do, however, have the advantage of being relatively comprehensive.

The Washington State Department of Ecology attempts to estimate the economic costs of fish kills by forecasting the costs of raising replacement fish in a state hatchery.[8] One limitation to this program is its failure to make estimates for species that, because they provide no direct human benefit, are not raised in hatcheries. Such species may be important to the viability of an ecosystem that, in its entirety, now benefits or in the future will benefit humans.

Despite these isolated attempts to assign economic values to wildlife, serious difficulties remain in attempting responsibly (*a*) to estimate the amounts of damage caused to different wildlife species by nonpoint-source pollutants and (*b*) to place appropriate values on the different

Figure 5.1
Values Assigned to Freshwater Fish by State of Florida

Common name	Size	Value
Alewife	all	$ .30
Bait fish	all	.03
Bass		
Chipola, Shoal	0–6″	.50
	7–12″	1.75
	over 12″	2.50 lb.
	over 3 lb.	10.00 lb.
Largemouth	0–6″	.50
	7–12″	1.75
	over 12″	2.50 lb.
	over 8 lb.	10.00 lb.
Striped	all	10.00
Suwannee	0–6″	.50
	7–12″	1.75
	over 12″	2.50 lb.
	over 3 lb.	10.00 lb.
Bowfin, Mudfish	0–6″	.03
	7–13″	.10
	over 13″	.25 lb.
Bullhead		
Brown, Speckled	0–6″	.10
	7–12″	.35
	over 12″	.40 lb.
Flat	all	.10
Yellow (Cat)	0–6″	.10
	7–12″	.35
	over 12″	.40 lb.
Carp	all	.25 lb.

Common name	Size	Value
Catfish		
Channel	0–7″	.10
	8–14″	.50 lb.
	over 14″	1.00 lb.
White	0–7″	.10
	8–14″	.50
	over 14″	1.00 lb.
Crappie, Speck	0–6″	.30
	7–12″	1.25
	over 12″	2.50 lb.
	over 2 lb.	5.00 lb.
Darter	all	.10
Eel (American)	all	.25
Gar	0–8″	.05
	9–16″	.15
	over 16″	.25 lb.
Alligator gar	all	2.00 lb.
Madtoms	all	.10 lb.
Mullet		
Black, Striped	all	2.00
Silver, White	all	2.00
Pickerel		
Chain, Jack fish	0–7″	.25
	8–14″	1.00
	over 14″	3.50 lb.
	over 4 lb.	5.00 lb.
Redfin	all	.25
Pirate perch	all	.10
Shad		
American	all	7.00
Gizzard	all	.05
Threadfin	all	.05
Snook,	all	20.00

Common name	Size	Value
Sucker	all	.20
Sturgeon	all	50.00 lb.
Sunfish (Panfish)		
Banded	all	.25
Banded pigmy	all	.25
Blackbanded	all	.25
Bluegill, Bream	0–5″	.25
	6–10″	1.25
	over 10″	3.00 lb.
	over 1½ lb.	5.00 lb.
Bluespotted	all	.25
Dollar	all	.25
Everglades pigmy	all	.25
Flier	all	.25
Longear	all	.25
Mud (Perch)	all	.25
Redbreast	0–4″	.25
	5″–1 lb.	1.00
	over 1 lb.	5.00 lb.
Shellcracker		
(Redear)	0–5″	.25
	5–10″	1.25
	over 10″	3.00 lb.
	over 2 lb.	5.00 lb.
Southern rock bass	all	.40
Stump-knocker	all	.40
Warmouth	all	.40
Tilapia	all	.25
All fish not specified		.10

Source: State of Florida, Department of Pollution Control, "Assessment of Damages," chapter 17–11 in *Rules of State of Florida*, adopted May 15, 1971.

species even if the damage can be estimated. As a result, this study places no value directly on biological impacts, although some of these impacts are reflected in the estimates given below for recreational and commercial fishing damages. This gap is probably the most significant in the series of estimates provided in this report.

Recreation

The economic costs of degraded recreational opportunities can be experienced in various forms. Dirty water makes fishing more difficult and swimming, boating, and beach going less pleasant. Some recreationists may forgo a recreational experience entirely. Some may shift to an alternative, less degraded facility, spending additional time and resources to do so. Efforts may be made to compensate for a loss by building an alternative recreational facility such as a swimming pool, by artificially stocking a water body with desirable game fish, by building siltation traps upstream of a recreational site, or by investing in a more attractive complementary facility (for instance, a fancier bath house or special wading pool). Usually, however, the damages are endured or the recreational opportunities are sought elsewhere.

Recreation has become a major economic activity in the United States, with Americans spending more and more time and money on leisure activities (figure 5.2). Consumer expenditures on recreation increased from $58 billion in 1965 ($140 in 1980 dollars) to $244 billion in 1981 ($220 in 1980 dollars), a 57 percent increase in constant dollars.[9]* In 1979, over 6 percent of disposable income was spent on recreation nationwide, and leisure activities accounted for 1 out of every 15 jobs in the United States.[10] In some regions of the country, these percentages have been much higher. For instance, a recent study in Pennsylvania showed that about $11.8 billion, about 12 percent of disposable income, was spend by consumers on leisure activities and accounted for approximately 400,000 jobs.[11] Sport fishing is big business in many local economies: salmon and trout in the Northwest and Northeast, walleye and perch in Lake Erie, and bass in the South.[12]

Various analyses have attempted to estimate the value of recreational opportunities and how they are affected by pollution. A study of Lake Chicot in Arkansas concluded that soil erosion was causing as much as $9.6 million a year in recreational damages.[13] Interviews of visitors

---

*The consumer price index was used to convert these dollar figures to their 1980 equivalents.

Figure 5.2
Recreational Activities by Americans in 1982

Activity	Participation reported in 1982 survey (percentage of survey respondents)
Boating	28
Canoeing or kayaking	8
Sailing	6
Motorboating	19
Other boating or watercraft sport	6
Waterskiing	9
Outdoor swimming	53
Swimming in an outdoor pool	43
Other outdoor swimming	32
Fishing	34
Camping	24
Birdwatching or other nature-study activities	12
Picnicking	48

Source: U.S. Department of the Interior, National Park Service, Recreation Resources Division, "Summary of Selected Findings" (April 1984 Draft). Final version published in National Park Service and U.S. Department of Commerce, Bureau of the Census, *1982–83 Nationwide Recreation Survey* (Washington, D.C.: U.S. Government Printing Office, 1985).

at a Michigan lake indicated $15,600 a year in damages, although a statistical analysis showed no significant relationship between recreational use and water quality.[14] Recreational benefits were found to be important in 25 of 28 lake-cleanup projects analyzed for the U.S. Environmental Protection Agency (EPA).[15] Controlling erosion on federal grazing lands in 10 western states was projected to create recreational benefits of about $78 million ($110 million in 1980 dollars).[16] None of these studies, however, provides a basis for making a national damage-cost estimate.

A 1982 study by Resources for the Future (RFF) does provide such a basis.[17] This study began by estimating the value of different types of fishing and how the total amount of fishing would be affected by improved water quality. For instance, it found that the value per fishing day is at least 50 percent higher when angling for cold-water game fish than when angling for catfish or other "rough" fish (figure 5.3). Based on an extensive survey, the study also found that, as water quality improves, "more people become anglers, and each angler is projected, on the average, to spend more days per year fishing (with a few exceptions in the trout category)."[18] The study placed "the total value of freshwater recreational fishing (excluding Great Lakes fishing) [at] between $12 and $27 billion per year, depending on the assumptions governing the per day values.''[19]

The study then estimated how much this total value would increase as a result of pollution-control efforts. It concluded that implementing all planned point-source controls (including the best available technology controls imposed on industry), neutralizing acid mine drainages, and controlling nonpoint-source pollution from cropland probably would create total freshwater recreational fishing benefits of $162 to $966 million per year. The benefits resulting from the first stage of cleaning up pollution from point sources would "probably lie between $100 million and $700 million per year.''[20] Eliminating all of the sediment discharges from nonirrigated cropland, it was estimated, would produce another $22 to $105 million in benefits—or 13 to 17 percent of the total benefits.*

Another recent study, by economist A. Myrick Freeman III, reviewed and analyzed several efforts to assess the benefits of water pollution control to provide benefit estimates for five different types of water-based recreation experiences.[23] The range of benefits for all five categories was equivalent to $2.0 to $10.5 billion in 1980 dollars with a single-value estimate of $5.5 billion (figure 5.4).[24] The study, however, did not indicate what portion of these benefits was attributable to erosion control.

To adjust for this shortcoming, this chapter first uses Freeman's estimates as the basis for estimating the costs to recreation caused by water pollution and then uses the RFF study to determine what proportion of those costs is attributable to sediment.† The RFF study,

---

*These estimates all were based on 1975 population and per capita income and, therefore, would presumably be higher when applied to 1980 conditions.[21] The benefits resulting from reducing cropland sediment were estimated on the assumption that "each pound of organic carbon in nonpoint source sediment is equivalent to 0.1 pound of biochemical oxygen demand.''[22] Thus, no direct account was taken of the impacts of suspended sediment, sedimentation, nutrients, pesticides, or other associated contaminants other than organic compounds. Presumably, including these other impacts would have increased the benefits associated with sediment control. However, no sediment control measure is likely to totally eliminate sediment loadings.

†As an alternative to this approach, nonpoint sources could be allocated an amount of recreation benefits proportional to their contribution to total conventional pollutant loadings. This assumption would have resulted in the recreational damages being two to three times higher than those given here.

Figure 5.3
Fishing Values

Type of angling	Value per day (1978 dollars)		
Cold-water game fish	$10.96	to	24.09
Warm-water game fish/panfish	$9.65	to	21.43
Catfish/other rough fish	$7.00	to	16.03

Source: William J. Vaughan and Clifford S. Russell, *Freshwater Recreational Fishing: The National Benefits of Water Pollution Control* (Baltimore: Johns Hopkins University Press, 1982), p. 148.

Figure 5.4
National Recreational Benefits Resulting from Water Pollution Control (billion dollars)

	1978 dollars		1980 dollars*	
Activity	Range of estimates	Single-value estimate	Range of estimates	Single-value estimate
Freshwater fishing	0.5–1.4	1.0	0.5–1.7	1.2
Marine sports fishing	0.1–3.0	1.0	0.1–3.6	1.2
Boating	1.0–2.0	1.5	1.2–2.4	1.8
Swimming	0.2–2.0	1.0	0.2–2.4	1.2
Waterfowl hunting	0.0–0.3	0.1	0–0.4	0.1
Total	**1.8-8.7**	**4.6**	**2.0-10.5**	**5.5**

* Conversion to 1980 dollars using gross-national-product deflator

Source: A. Myrick Freeman III, *Air and Water Pollution Control: A Benefit-Cost Assessment* (New York: John Wiley & Sons, 1982), p. 161.

however, considered only cropland sediment. This estimate is converted here to include all sources of eroded sediment by multiplying it by 2.5 to 3.0,[25] providing a total single-value cost estimate of $2 billion in 1980 dollars (figure 5.5). Of this total, the largest impact is on boating—$680 million.

None of these estimates includes possible accidents or deaths that swimmers and boaters may experience because of turbid water. Chapter 3 indicated that there could be as many as 200 such deaths a year. Assigning a value to human life is a particularly controversial question in estimating environmental damages. However, assuming a value

Figure 5.5
Recreational Damages
(million 1980 dollars)

Type of impact	Range of estimates	Single-value estimate	Cropland's share
Freshwater fishing	160- 870	450	180
Marine fishing	330-1,800	450	180
Boating	390-1,200	680	270
Swimming	70-1,200	450	180
Waterfowl hunting	0- 200	50	20
Accidents—fatal	0- 200	–	–
—nonfatal	0- 100	–	–
Total (rounded)	**950-5,600**	**2,000**	**830**

Source: Original Conservation Foundation research.

of $1 million per life indicates that such deaths might cost the United States as much as $200 million dollars a year.* Arbitrarily increasing this by 50 percent to take account of accidents that do not result in death would give a value of $300 million dollars. However, no estimate has been included for either of these costs.

There is also a question of indirect costs. The economic consequences of the direct damages can be experienced by the many sectors of the economy that service recreationists, although no agreement exists among economists on whether and how such secondary costs should be valued. One study of the impact of turbidity and sediment on fishing in an Illinois lake summarized these impacts as follows:

> At one time, special trains brought sport fishermen to towns, such as Havana, along the Illinois River, and freight trains hauled away commercial fish to Chicago or New York. Now, Dixon's fee fishing area in Peoria imports carp from Wisconsin, the restaurants along the Illinois River buy channel catfish from Arkansas, and the residents of beach communities, such as Quiver Beach and Baldwin Beach, in Mason County, are no longer able to swim in the bottomland lakes or even to launch their boats from their cottages in midsummer.[27]

Thus, the total economic costs associated with downstream recreational impacts may extend well beyond the costs experienced by recreationists themselves. However, these "secondary" or "indirect" costs

*EPA currently uses a value of $400,000 to $7 million per life saved in conducting cost-benefit analyses of proposed regulations.[26]

initially imposed by erosion will be greater than the benefits that could be expected from erosion control. Such costs occur when a recreational site is first damaged. But society adjusts to these problems, and new services become available to serve the recreationists at alternative sites. If the originally degraded site is cleaned up, no indirect savings will be realized, because the services will already have been provided elsewhere. In some cases, it may even be argued that cleaning up an original site will create indirect costs because the services at the alternative site will suffer and new ones will have to be provided again at the original site.

Water Storage in Lakes and Reservoirs

As indicated in chapters 3 and 4, erosion can impose three major types of costs on water-storage facilities. One is sedimentation in lakes and reservoirs filling up valuable storage capacity. A second is the way in which sediment and associated contaminants may modify evaporation and transpiration rates. The third is the way that pollution can keep lakes and reservoirs from serving their intended functions.

Lost Storage Capacity

There are four basic ways to respond to the problem of lost storage capacity caused by sedimentation. The first is to attempt to anticipate the loss and build extra capacity to collect the sediment. The second is to try to prevent the sediment from settling in a reservoir. In addition, sediment can be removed from a reservoir by dredging. Finally, lost original capacity can be compensated for by building replacement facilities or finding other sources of water.

Previous efforts to estimate these costs have indicated that they may be quite substantial. For instance, a 1947 study, based on a survey of 600 reservoirs, estimated an average annual cost of $50 million (about $180 million in 1980 prices).[28] A 1976 estimate placed the cost at $100 million (about $140 million in 1980 dollars).[29] A 1979 study estimated that the reservoir-capacity-reduction costs in Illinois alone were approximately $20 million.[30]

These costs probably are increasing continually. One reason is that the cost of constructing new storage capacity has risen substantially in recent years. Construction costs in general have increased rapidly, as evidenced by the U.S. Bureau of Reclamation's construction cost index, which rose by 120 percent from 1971 to 1980, 34 percentage points more than the gross-national-product deflator.[31]

Probably even more important is the fact that most of the good dam sites—deep, narrow valleys downstream of large basins—have been used up. Remaining sites typically require a longer and higher dam in order to provide a given amount of capacity. As one analyst observed, "sites for low-cost dams are disappearing and will soon be gone."[32] Largely as a result, the average amount of construction material used in a dam to provide one acre-foot of capacity was seven times greater during the 1960s than during the 1930s (figure 5.6).*

Some of this expensive capacity serves only to store sediment. In designing most reservoirs, engineers first determine the capacity required to provide the primary outputs of the reservoir and then add sufficient additional capacity, called the "sediment pool," to store the expected sediment that will be trapped in the reservoir during its operating life.[33] In areas experiencing high erosion rates, such as Missouri, the sediment pool can take up to 30 percent of the total capacity.[34] Eighteen dams proposed for the upper Little Arkansas River watershed in Kansas would have 15 percent to 30 percent—averaging 23 percent—of their capacity allocated to sediment storage.[35] For watershed projects planned by the U.S. Department of Agriculture (USDA), an average of 15 to 20 percent of capacity is allocated to sediment storage.[36]

Approximately 5 million acre-feet a year of new usable reservoir capacity was built in the United States during the 1970s.[37] Assuming that the additional unusable capacity allocated for sediment storage averaged 15 to 25 percent of this and that the 1970s rate of dam construction is continuing, approximately 750,000 to 1.25 million acre-feet of capacity a year are being constructed for sediment. Reviewing actual and estimated construction costs for several new dams indicates that new reservoir capacity can be expected to cost $300 to $700 per acre-foot or more (figure 5.7). Using this cost range indicates that the amount being spent to construct reservoir capacity to store sediment is from $220 to $880 million per year.

The second approach to dealing with sediment is to attempt to prevent it from ever settling in a reservoir. Efforts have been made to reduce the amount of sedimentation by placing outlets at the bottom

---

*However, another reason for this increase is the fact that many earlier dams were made out of concrete or masonry, whereas most recent dams have been made out of earth. The former require less material than the latter. But a comparison of "earthfill" dams alone would undoubtedly show a similar, if less dramatic, trend.

Figure 5.6
Volume of Dam Material Required Per Acre-Foot of Reservoir Capacity, pre-1920 through 1960s

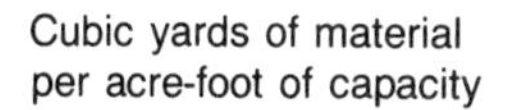

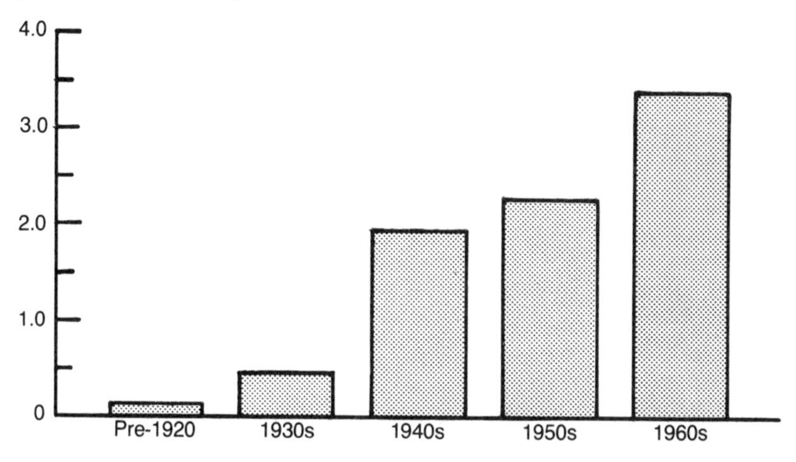

Source: Walter B. Langbein, "Dams, Reservoirs, and Withdrawals for Water Supply—Historic Trends" (Open File Report 82-256 by the U.S. Geological Survey, Reston, Va., 1982).

of dams in the hope that they will draw the sediment off from reservoir bottoms or by constructing baffles that attempt to deflect sediment from reservoir bottoms up to the spillways, where it will flow on downstream.[38] It is unclear, however, how effective these techniques are, and none of them has demonstrated success in redmoving sediment that has already settled. They also suffer the disadvantage of transferring the sedimentation problem downstream. There seems to be very little information about the costs of such techniques, and their use appears to be very limited.

In some cases, if the sedimentation is severely affecting reservoir capacity or causing other unwanted problems, it can be removed by dredging or excavation.[39] The Baltimore District of the U.S. Army Corps of Engineers removed 18,000 cubic yards of sediment from three small dams in 1980 as part of its normal maintenance program.[40] Los Angeles County removed over 1 million cubic yards of material from flood-control basins after the 1960–61 storm season.[41] Such dredging, however, usually is feasible only for small reservoirs since there may be no feasible way of disposing of the dredged sediment from larger reservoirs.[42]

Figure 5.7
Costs of Surface Storage Capacity

Storage site	Completion date	Cost at construction (dollars per acre-foot)	Adjusted cost (1980 dollars per acre-foot)
Reudo Dam-Arkansas Fryingpan Project[1]	1963	230	620
Dallas Creek Project, Colorado[1]	1976	595	819
Leroux Creek, Colorado Grand Mesa Project[1]	1978	212	260
Slip-up Creek, Sioux Falls, South Dakota[1]	1978	275	338
Tyzack Creek, Central Utah Project Utah[1]	1974	957	1,649
Dry Creek, Rice County, Kansas[2]	1981	701*	642
Lone Tree Creek, McPherson County, Kansas[2]	1981	624*	571
Sand Creek, Rice County, Kansas[2]	1981	575*	526
Unnamed Creek, Rice County, Kansas[2]	1981	540*	494
North Branch Horse Creek, Rice County, Kansas[2]	1981	751*	687
North Fork, Little Arkansas River, Rice County, Kansas[2]	1981	537*	524
Cedar Run Impoundment, Prince William County, Virginia[3]	1978	896*	1,100

* Estimated Costs

Sources:
1. U.S. Department of the Interior, Planning Reports and Environmental Impact Statements
2. U.S. Army Corps of Engineers, *Upper Little Arkansas River Watershed, Kansas, Survey Report and Environmental Impact Statement, Vol. I*, Oct. 1981.
3. U.S. Army Corps of Engineers, *Draft-Metropolitan Washington Area Water Supply Study, Appendix I*, March 1983.

Dredging also typically is much more expensive than building new storage capacity. A 1978 study of reservoir sedimentation in Illinois estimated that the minimum cost of dredging material was $1.25 per cubic yard ($1.63 per cubic meter, or $2.00 per cubic meter in 1980 dollars).[43] The estimated costs of hydraulic dredging undertaken during the mid-1970s as part of EPA's Lake Restoration Program varied from $0.27 to $3.16 per cubic meter.[44] And the costs can rise much higher, particularly if it is difficult to identify a suitable disposal site for the dredged material.[45] At a cost of $2.00 per cubic meter, dredging costs approximately $2,500 per acre-foot of material removed—3 to 8 times the cost of new capacity estimated earlier.*

Since the passage of the 1972 amendments to the Federal Water Pollution Control Act, many of the most environmentally damaging methods for disposing of dredged material have been prohibited. The result has been a significant increase in dredging costs. For instance, after it cost $1.2 million to dredge the Illinois River in 1975, an estimated additional $1 million had to be spent to properly dispose of the dredged materials.[47] A series of case studies by the General Accounting Office indicated that less-damaging disposal methods still could increase project costs by 65 to over 1,000 percent.[48] And, even with the most expensive disposal methods, there would still be some environmental damage and aesthetic cost.†

Few researchers have attempted to estimate the amount of money spent on lake and reservoir dredging nationally. Reports from different sections of the country certainly indicate that these costs are significant. The state of Illinois estimated that, by 1978, it was spending $17.7 million annually to dredge lakes.[50] By 1980, San Francisco was spending an average of a quarter of a million dollars annually to remove sediment from 3 lakes near the city.[51] Almost $20 million was spent dredging sediment from 15 lakes receiving funds from EPA under its Clean Lakes Program, although these projects were not all undertaken in one year.[52] On the basis of such episodic evidence, it can be inferred that the nation is probably spending at least $20 to $100 million per year on lake and reservoir dredging. For this report, the point estimate is assumed to be $50 million per year, with $15 million of this attributable to cropland erosion.

---

*However, excavating dry materials with earth-moving equipment from flood-control check dams is usually less expensive than dredging, though removing wet materials is not.[46]

†Disposing of dredged material can sometimes create economic benefits, however. For instance, some of the material dredged for a proposed deep-water access to the ports of New Orleans and Baton Rouge would be used to create 11,600 acres of marsh.[49] At a value of $1,000 per acre, this marsh would have a capital value of $11 million or an annualized value of about $1 million per year.

The final type of cost is the value of the capacity in existing reservoirs that is being lost through sedimentation. Most likely, this lost capacity will be replaced eventually—either with new capacity provided by a substitute structure in the same watershed or by importing water from more-distant ground- or surface-water sources.* Chapter 3 estimated that 1.4 to 1.5 million acre-feet of capacity is consumed each year. At a cost of $300 to $700 per acre-foot of capacity, the replacement cost of this capacity would be $420 million to $1.05 billion per year.

However, the actual economic cost of this lost capacity is less than this amount since much of the sedimentation causing the loss is occurring in sedimentation pools. There is no cost to this capacity loss until these pools are filled and additional sedimentation begins to fill reservoir space allocated to other purposes. Moreover, even when sediment is filling usable capacity, the effects of the decreased storage may not be experienced for several years because reservoirs usually are constructed to meet future, not present, demand.

Thus, replacement costs for lost capacity must be "discounted" to reflect the fact that they will not be experienced until sometime in the future. Doing so indicates that the "present value" of the future replacement costs lies between $60 and $600 million a year.[53] It should be understood, however, that these assumptions ignore the likelihood that future unit costs for replacement capacity will be much higher than the current cost of constructing that capacity because the available sites will, for the most part, be even less desirable.

Evaporation

Impacts on evaporation from reservoirs can be particularly important because, by the time evaporation occurs, all the costs of making the water available and of operating and maintaining the water-storage facility have been paid. Providing replacement water is likely to be very expensive. A recent estimate of the cost of making additional amounts of surface water available for southern California suggests that they could be as high as $180 per acre-foot.[54] A 1981 analysis of the Geary project in Oklahoma found that the water it would make available would cost $187 to $255 per acre-foot.[55] These costs, however,

---

*The lost capacity might not be replaced in areas where water use is decreasing, but such areas are uncommon.

probably are higher than the norm and therefore not appropriate for calculating a national average cost of new water supplies. The estimates given here assume a range of $50 to $150 per acre-foot.

As indicated in chapter 3, there is very little information on which to base estimates of changes in evaporation. Scattered previous studies have indicated that as much as 1 million acre-feet a year of increased evapotranspiration resulting from vegetation growth and increased reservoir surface area is possible.* Counteracting these impacts, however, may be the reduction in surface-water temperature that suspended sediment can cause. Presumably, this same effect could result from extensive algal growth caused by nutrient loadings.

The dilemma is that these impacts could be quite significant, but there is very little information to support an estimate of their magnitude. The estimates in figure 5.8 are based on two entirely arbitrary assumptions: (*a*) the amount of increased evapotranspiration is somewhere between 100,000 and 1 million acre-feet per year; and (*b*) the amount of water saved through reduced evaporation is twice as great as the amount lost in increased evapotranspiration.† These two assumptions result in ranges of $5 million to $100 million of costs for increased evapotranspiration and $10 million to $200 million in benefits from reduced evaporation. However, both estimates could be off by an order of magnitude. The potential uncertainty associated with these estimates is probably greater than most of the others included in this chapter.

Water-Quality Improvement

The final type of water-storage cost associated with erosion is the correction of problems caused by nutrients and other contaminants in lakes and reservoirs. Possible responses to these problems include: (*a*) accepting the taste, odor, and color problems that contamination can cause, (*b*) increasing the amount and sophistication of the water treatment that occurs before the water is used, (*c*) developing an alternative

*Evapotranspiration includes both evaporation directly from water bodies and transpiration of water by vegetation.

†The evaporation benefits created by suspended sediment and algal growth may not be as important as estimated because evaporation losses are most serious in large western reservoirs where these conditions may be less prevalent. Much of the suspended sediment settles out of the surface soon after entering the reservoir, and algal growth is not as extreme as it can be in smaller reservoirs.

Figure 5.8
Damages to Water-Storage Facilities
(million 1980 dollars)

Type of impact	Range of estimates	Single-value estimate	Cropland's share
Construction of sediment pools	220-880	350	110
Dredging and excavating	20-100	50	15
Replacement capacity	60-600	300	100
Reduced evaporation*	(−)10-(−)200	(−)100	(−)30
Increased evapotranspiration	5-100	50	15
Water-quality treatment	17- 85	40	15
Total (rounded)	**310-1,600**	**690**	**220**

*Reduced evaporation is a benefit provided by suspended sediment and therefore is included as a negative number in this table of damages.

Source: Original Conservation Foundation research.

water-supply source (groundwater or surface water) that does not suffer from such a problem, and (*d*) treating the source of supply to eliminate the problem. The first three of these responses are discussed later in this chapter's section on water-treatment facilities. The fourth is considered here.

In recent years—perhaps in response to the continued degradation of water quality in lakes and reservoirs—interest has increased in treating water bodies to eliminate water-quality problems. This change in approach was reflected, for instance, in the way the state of Connecticut responded to problems in the Trap Falls Reservoir. After the initial response—increased pumping from an alternative source (wells)—was found to be "a very expensive and not entirely successful solution,"[56] a series of projects was undertaken to improve the quality of the reservoir's water and to treat it more adequately before it was used. Polymer coagulants were added to water flowing into the reservoir, and dikes were built to increase the reservoir's water-retention time (at a 1973 cost of $200,000), promoting increased sedimentation of suspended materials. A water-filtration plant was also constructed.

Many communities have had to dredge contaminants from lake bottoms to control eutrophication. For instance, a Massachusetts lake was dredged in the mid-1970s in an effort to eliminate nutrients from sewage and runoff.[57] Those nutrients had stimulated the growth of algae and other aquatic plants, making the lake useless for boating, fishing, and swimming. In addition, sedimentation had made the lake shallow, further encouraging the growth of submerged and floating aquatic plants.

Restoration of the lake, which also included a watershed management program to reduce future nutrient runoff, took three and a half years and cost over half a million dollars.

Despite such expense, lake restoration efforts are becoming increasingly common as satisfactory unused reservoir sites become scarce and people become more aware of the potential damages of pollutants accumulating in their water supply. Dredging, however, is not the only means of lake restoration. Copper sulfate is frequently added to water to kill algae. Other chemicals, such as aluminum sulfate, are often used to bind phosphorus in the sediment, thereby preventing nutrients from returning to the water and stimulating plant growth.[58] Some water jurisdictions have resorted to draining unused old reservoirs and burning the highly organic bottom sediments.[59] Finally, "biological" control measures have sometimes been tried. For instance, the Chinese grass carp was introduced in Arkansas and Florida waterways to control Eurasian milfoil and other plants.[60]

Because of the widespread concern about water quality in lakes and reservoirs, Congress created a Clean Lakes Program in the 1972 amendment to the Federal Water Pollution Control Act.[61] Under this program, EPA has sponsored a series of research and demonstration projects to evaluate the feasibility of restoring degraded lakes.[62] By the end of 1983, the agency had awarded $70.6 million in grants to cover half of the costs of restoring 127 lakes, equivalent to a total cleanup cost of over $1 million per lake.[63] According to a study of 28 of these projects, most of the grants have been for between $100,000 and $500,000 (figure 5.9). They have been used to help finance dredging, physical and chemical/biological treatment of lake sediment, artificial aeration, chemical treatment and dilution of lake waters, and construction to divert inflow from storm sewers and other pollution sources.

A 1981 legislative audit of Wisconsin's Inland Lake Renewal Program found that $15.3 million were spent on 25 projects between 1974 and 1981 and that $2.8 million were spent in 1981 alone.[64] The state supplied 25 percent of these funds, the federal government 27 percent, and local governments 47 percent. By 1981, however, only $300,000 of the $2.8 million came from the federal government, and $1.04 million came from the state. These data do not include cleanup projects supported only by local and private funds.

If every state were as active as Wisconsin, that state's $2.8 million would be equivalent to national expenditures on lake cleanup in the range of $100 million a year. However, because Wisconsin has an unusually active lake improvement program, this $100 million estimate

Figure 5.9
EPA Clean Lakes Program

Lake	Restoration measures: Dredging	Physical and chemical treatment of lake sediment	Artificial aeration	Chemical/biological, and dilution of lake waters	Construction—divert sewer/stormwater	Construction—inflow protection, banks, other	Summary of benefits: Recreation	Aesthetics	Flood control	Economic development	Fish/wildlife	Agriculture	Property value	Public health	Water supply	Education/research and development	Miscellaneous	Total discounted benefits* (thousand dollars)	314 grant amount (thousand dollars)
Annabessacook				X			X	X		$		X	$			X		23,246.1	497.906
Bomoseeh		X					$	X								X	X	1830.5	74.64
Buckingham	X	X					$			X						X		127.7	23.25
Charles			X		X		$	$								X		2286.6	387.163
Clear					X	X	$	$		X	X			X		X		471.5	358.682
Cochrane						X	$	X			$		X			X		52.5	9.906
Collins Park	X					X	$	X								X		51.7	79.355
Ellis	X	X			X	X	$	$	$	$	X	$		X		X	$	11123.0	1625.0
Fifty-Ninth Street	X	X			X	X	$	X	X					X				4837.0	498.035
Frank Holten	X					X	$	$		X				X		X	X	1862.3	927.0
Henry	X					X	$					X	X			X	$	134.2	220.0
Jackson						X	$	$			X		X					7309.8	725.663
Lansing	X						$			X			X			X		1155.9	800.0
Liberty	X			X	X	X	$	$					X	X		X		813.0	577.975

Lilly	X						X	X			X		$			X		2880.0	350.0
Little Pond				X				X							$	X		212.2	9.946
Loch Raven			X				$	$							$	X		11944.1	150.9
Medical				X			$	X			X		$					931.7	128.217
Mirrow and Shadow			X	X	X		$						X	X		X		312.7	215.0
Moses				X			$	X								X	X	534.7	3251.0
Nutting	X						$	$		$			X	X		X		5292.1	241.159
Penn			X	X		X	$	$		X	X					X	$	186.0	87.9
Rivanna			X				$								$	X		923.5	63.835
Steinmetz		X		X	X		$	X	X	X				$		X		126.3	36.68
Temescal	X					X	$	$						$		X	$	1112.5	315.618
Tivoli	X				X	X	$	X	X	X				X		X		240.1	202.645
Vancouver	X					X	$	X		X		$		X		X		47370.0	3468.328
Washington Park	X						$	$								X		120.8	23.25

$ — Benefit measured in monetary terms
X — Benefit measured in qualitative terms

* Total discounted benefits include "$" items only.

Source: JACA Corporation, *An Assessment of Economic Benefits of 28 Projects in the Section 314 Clean Lakes Program*, Report prepared for the U.S. Environmental Protection Agency, Office of Water Planning and Standards (Fort Washington, Pa.: JACA Corporation, 1980).

can be assumed to define the top of the expenditure range. The bottom of the range is defined by only those projects supported in part by EPA grants—approximately $20 million per year. Nationally, about 85 percent of nutrient loadings are estimated to come from nonpoint sources, so between $17 and $85 million of the costs can be attributed to them.[65]

Total Economic Cost

Taking all the various sediment-associated impacts on water-storage facilities into account, the total economic costs range from $310 million to $1.6 billion dollars, with a single-value estimate of $690 million dollars (figure 5.8). Cropland's share would be approximately $220 million dollars.

Navigation

The major effects of sedimentation in channels and harbors—increased shipping accidents, reduced capacity of harbors and navigation channels to handle boats and ships, shipping delays, and required dredging—all can have economic repercussions. In addition, suspended sediment, nutrients, and other pollutants may require increased expenditures for repairing and maintaining ship engines and other equipment.

Coast Guard records indicate that commercial shipping accidents in 1980 cost almost $300 million in damage to vessels, cargo, and other property.[66] Almost $60 million of those costs were due to groundings, 84 percent of which occurred in inland waters. In all, about two-thirds of all accidents occurred in inland waters, and about half of those involved collisions or groundings that could be linked to sedimentation. That is, they might be avoided if the channels were deeper and wider. On the basis of the Coast Guard data, sedimentation is estimated in this report to contribute $20 to $100 million dollars in commercial shipping damages, with a single-value estimate of $40 million (figure 5.10). These costs include neither the public health and environmental damages that result from leaking fuel and spilled cargoes nor the deaths or injuries suffered by people involved in the accidents.

It is not possible to estimate economic costs for the effects on navigation of reduced capacities in navigation channels and harbors—primarily, delays and forced usage of smaller vessels. Direct evidence linking sedimentation with such impacts is almost nonexistent. However, this does not mean that the costs are small. For instance, when silta-

Figure 5.10
Costs of Navigation Impacts
(million 1980 dollars)

Type of impact	Range of estimates	Single-value estimate	Cropland's share
Delays to commercial shipping		no estimate	
Accidents	20-100	40	13
Damage to engines		no estimate	
Dredging	400-700	520	170
Dredge spoil disposal		no estimate	
Total (rounded)	**420-800**	**560**	**180**

Source: Original Conservation Foundation research.

tion in the south pass of the Mississippi River below New Orleans following a 1974 flood required ships to wait extra periods of time and to travel to other ports, it undoubtedly caused real resources to be consumed and, therefore, imposed economic costs.* Similarly, although shallow channels undoubtedly force shipping companies to use smaller ships or lighter loads and damage ship's engines through sand abrasion, there is no reliable evidence to indicate the extent of these problems.

Probably the largest erosion-caused costs to navigation are the expenses of removing sediment from channels and harbors. A major—and rapidly increasing—portion of this cost, as mentioned earlier, is disposing of sediment after it has been dredged. For instance, the actual dredging of the rivers in Michigan cost only $2.39 per cubic yard in 1979, but proper disposal cost $5.29 per cubic yard.[68]

Some $305 million was spent by the Corps of Engineers in 1980 for maintenance dredging of harbors and waterways, including disposal costs.[69]† Even that amount was not adequate to maintain suitable channel depths throughout the year. Moreover, the Corps is only responsible for about half of all the dredging done in the United States; the

*This siltation was estimated to have "reduced foreign trade at the port of New Orleans by an estimated half billion dollars during the spring of 1974."[67] Most of this trade probably was diverted to other ports.

†The Corps's maintenance-dredging budget increased about 50 percent between 1980 and 1984, predominately as a result of increased disposal costs.[70]

remainder is carried out by other government agencies and private sources.[71] Thus, total dredging costs for the nation can be estimated to be between $400 and $700 million a year, with a single-value estimate of $520 million.

Finally, it is safe to assume that suspended sediment, algae, and other pollutants at least occasionally clog ship-engine cooling systems and cause them to overheat. However, almost no information exists on the extent of these damages or their economic costs.

Other In-Stream Impacts

Other in-stream impacts associated with the products of soil erosion include reductions in the population of commercial fisheries, in riparian property values, and in what economists refer to as "preservation values."*

There have been no analyses that estimate the economic costs specifically caused by sediment in any of these categories. However, one recent study estimated that removing conventional water pollutants from *all* sources would create annual benefits to commercial marine and freshwater fisheries of $800 million in 1978 dollars ($950 million dollars in 1980 dollars).[72] Using this total and assuming that 32 to 51 percent of the damages are attributable to erosion and associated contaminants (these are the same percentages used in calculating erosion's proportion of recreational fishing benefits), erosion costs commercial fisheries between $300 and $480 million a year in 1980 dollars, with a single-value estimate of $400 million (figure 5.11).

Numerous studies have shown that degraded water quality lowers riparian property values. A 1975 analysis estimated that water pollution abatement would cause property values in the United States to increase by $61.7 million by 1980 (about $88 million in 1980 dollars).[73] This estimate, however, probably did not assume any sediment control. Other studies have found that aesthetic benefits (measured in part by changes in property values) equal 0.5 to 1.5 times the value of recreational (or "user") benefits.[74] In a recent study of lakefront real estate along the Canadian shore of Lake Erie, algae-fouled beach frontage

---

*Preservation values represent people's desires to preserve an environmental amenity such as clean water beyond any direct benefit that the amenity may provide them. For instance, even if people do not use a lake for recreation, water supply, or any other purpose, they may prefer that the lake be kept clean to provide them with the option of using it if they or their descendants ever wish to.

Figure 5.11
Other In-Stream Damages
(million 1980 dollars)

Type of impact	Range of estimates	Single-value estimate	Cropland's share
Commercial fisheries	300- 480	400	140
Property values (aesthetics)		not estimated	
Preservation values	160-2,000	500	180
Total (rounded)	**460-2,500**	**900**	**320**

Source: Original Conservation Foundation research.

was found to have property values 15 to 20 percent below those of comparable clean shorefront.[75] Another study of recreational properties along St. Albans Bay (Lake Champlain) in Vermont indicated a similar relationship.[76] Although water quality is, of course, not the only determinant of property values (quality of view, drainage, and isolation from other development are some other factors), it does seem to have a measurable impact.

Some of these changes in property values may reflect damages considered earlier in this chapter. For instance, if a property is used only for recreational purposes, its reduced value would be expected to be a result of recreational damages and should not be counted a second time. Because of this, no attempts are made here to estimate the effects of erosion on property values. Still, some aesthetic damages probably are reflected in property values but not included in this study's other cost estimates. They are therefore included, as compensation, with other "nonuser" damage estimates given below.

Very few serious attempts have also been made to measure how water quality affects an area's preservation value. One recent study of the South Platte River in Colorado concluded that the waterway's preservation value was $62 million dollars a year, approximately twice the value of the river for current recreational use.[77] Other studies have found that "nonuser" benefits (such as preservation values) are about 0.5 to 1.4 times the magnitude of user benefits.[78]

A 1982 review estimated that the value of all nonuser benefits (including aesthetic) resulting from controlling all water pollution probably would range from $500 million to $4 billion dollars a year in 1978 dollars, with a best estimate of $1.2 billion dollars ($1.4 billion in 1980 dollars).[79] Assuming again that 32 to 51 percent of these benefits would be attributable to soil erosion, it is possible to estimate that annual in-stream damages to preservation values in the United States total between $160 million and $2 billion, with a single-value estimate of $500 million. Cropland alone accounts for an estimated $180 million in preservation-value damages (figure 5.11).

OFF-STREAM IMPACTS

Estimating economic costs often is conceptually easier with soil erosion's off-stream impacts than with its in-stream effects. Most of the off-stream impacts result in clear expenditures to correct or avoid problems or in reductions in economic production. Unfortunately, however, this simpler conceptual situation is offset by a substantial lack of data on how serious the actual off-stream effects are. Thus, in the end, cost estimates are no more certain for erosion's off-stream impacts than for its in-stream effects.

Flood Damages

The Water Resources Council estimated that typical flooding conditions would create flood damages in the United States totaling about $5.0 billion in 1980 (in 1980 dollars).[80] Of these damages, $1.7 billion would have occurred in urban areas, $2.3 billion on agricultural lands, and the remainder in other nonurban areas.[81] About $1.9 billion of the total damages would have occurred in upstream reaches, with 68 percent of these costs representing damage to agricultural lands.*

Although the basic fact that sediment is a major contributor to flood damages is often cited, apparently only two serious attempts at national cost estimates have been made. In 1947, the cost of sediment damage to crops, clearing sediment off roads and property, and property damage resulting from increased flood heights due to stream aggradation, was estimated to account for at least 20 percent of the nation's flood damages, or $20 million ($72 million in 1980 dollars).[83] This estimate did not include another $50 million annually ($180 million in 1980 dollars) of other flood-related damage to agricultural lands. A second assessment, conducted in 1964, estimated that damages caused by sediment amounted to $87 million ($213 million in 1980 dollars) in upstream sections of rivers alone.[84]

---

*Upstream damages in these estimates are those occurring on streams with a drainage area less than 250,000 acres (approximately 400 square miles).[82]

Some other studies have estimated flood damages in particular regions. A study of 24 drainage basins in 1953 concluded that erosion-control programs could prevent the destruction of about $5 million ($15 million in 1980 dollars*) of crops annually.[85] Another study of the Piedmont area in North Carolina concluded that 1 out of every 10 acres on the floodplain was affected by sedimentation, and half of those had suffered at least a 20 percent reduction in productivity.[86]

Only one of these analyses appears to take account of the increased flood heights caused by stream aggradation. This omission is largely due to there being no reliable data about how extensive such aggradation is in the United States, although there is substantial evidence that it has occurred.

The most serious aggradation would be expected to occur in the upstream segments of river basins. But, as pointed out in chapter 3, upstream flooding may reduce peak flood flows downstream, reducing damages there. Since no information exists on the magnitude of either effect, an arbitrary assumption that the net effect can range from 0 to 10 percent of the upstream costs, or $0 to $190 million per year, is made here to give a rough order of magnitude for the significance of these impacts.†

The magnitude of the damages caused by the increased flood volumes that suspended sediment produces can be roughly estimated by considering together three separate sets of data: (*a*) the effect that sediment has on floodwater volume (shown in chapter 3, figure 3.9), (*b*) U.S. Geological Survey (USGS) measurements of the maximum sediment concentrations in the nation's 21 water-resource regions (shown in chapter 1, figure 1.3), and (*c*) Water Resources Council estimates of flood damages in each of those regions.[87]‡ If it is assumed that

*The index used here was the producer price index for foodstuffs and feedstuffs.

†The estimates in this section probably substantially overstate the extent to which flood damages would be decreased if cropland erosion were to be controlled. The floodwaters would continue to pick up heavy sediment loads from stream channels and from floodplains even if no sediment from eroding cropland were added to the streams. In fact, sediment-free floodwater could cause substantial damage by creating serious gully erosion along the streams and floodplains.

‡Because the sediment readings usually were taken no more frequently than once a month, it is likely that they understate the actual maximum concentrations occuring during floods.

the responsibility of sediment for flood damage in each region is directly proportional to the amount it increases the volume of floodwater, it can be concluded that a single-value estimate of $23 million in flood damages was associated with sediment-increased water volumes in 1980. For this study, it is arbitrarily assumed that this represented a point in a range of $10 million to $50 million.

Further economic costs occur after eroded sediment settles out of floodwater onto agricultural or urban lands. As noted in chapter 3, a major component of urban flood damage is the settling of sediment in streets, houses, and elsewhere.[88] A study of upstream flood damages concluded that 8.6 percent of total damages resulted from sedimentation.[89]* If it is assumed that 15 to 30 percent of the urban damages and 5 to 10 percent of other flood damages are caused by sedimentation, the total annual cost of sediment settling on land can be estimated at between $420 and $840 million.

When the cost estimates associated with flood heights raised by channel aggradation, flood volumes increased by suspended sediment, and the settling of sediment onto land are combined, they represent 9 to 22 percent of total flood damages in the United States. All of these estimates, however, refer only to property damage. Despite billions of dollars of investment and protection works, floods can also result in the loss of hundreds of lives. The average number of lives lost in floods during the 1970s was 176 per year (varying from a low of 74 in 1971 to a high of 540 in 1972).[90] Conservatively assuming that an average of 150 lives are lost annually in floods, and that sediment is responsible for the same proportion of lives lost as it is for flood damages, then sediment would be responsible for 14 to 33 deaths per year. Valuing these at $1 million per life indicates additional flood damage costs of $14 to $33 million.

Several other economic costs are associated directly or indirectly with sedimentation. One is the long-term loss in productivity associated with sedimentation of relatively infertile material on good agricultural land. About 12 percent of the nation's total cropland and 16 percent of its prime farmland is in floodplains, and thus subject to this effect.[91] The previously mentioned study of floodplains in the Piedmont area suggested that flood-related sedimentation may have cut the average productivity of all floodplain land in that region by around 2 percent. Assuming that this is the maximum likely loss, and that bottomlands

*This estimate may also include some costs associated with sedimentation in reservoirs.

account for the same proportion of crop income as they do of total cropland, the costs of reduced floodplain yields in 1980 could have been between $0 and $170 million.*

In addition, the aggradation of stream channels has flooded some formerly highly productive riparian farmland, making it completely unsuitable for agriculture.† On the other hand, these expanded wetlands may provide valuable wildlife areas, at least partially offsetting the destruction of agricultural land. There is no statistical data on the extent of either effect, however, so no estimate is made here of their magnitude.

Still, when all the more quantifiable damages are added together, it is possible to estimate total sediment-related flood damages at between $440 million and $1.3 billion (in 1980 dollars), with a single-value estimate of $770 million. Cropland's share of this would be approximately $250 million (figure 5.12).

Water Conveyance

Sediment can cause damages as it clogs drainage ditches before water actually reaches a waterway and canals and channels conveying water to a point of off-stream use or as it increases the costs of pumping water for off-stream uses.

Drainage Ditches

Much of the soil eroded from fields or construction sites never reaches a waterway, being deposited somewhere en route. Some of this deposition takes place in drainage ditches, where, if the material is not removed, it can cause localized flooding.‡

To prevent such damages, state and local highway departments spend a substantial portion of their budgets removing sediment from drainage ditches. A 1977 survey of Illinois road maintenance departments (both state and county) concluded that, each year, the removal of 2.5 million cubic yards of sediment from roadside ditches in the state costs $6.3

---

*Gross farm income from crops was $72.7 billion dollars in 1980.[92]

†This swamping of riparian land forces farmers farther up hillsides, where their activities may increase erosion and thereby further exacerbate sediment problems.

‡The costs of this flooding are not included in the flood costs indicated in the previous section.

Figure 5.12
Sediment-Related Flood Damages
(million dollars)

Type of impact	Range of estimates	Single-value estimate	Cropland's share
Increased flood heights from channel aggradation	0-190	50	16
Increased flood volumes	10- 50	23	8
Direct sediment damages			
Urban	260-510	350	110
Other	160-330	250	82
Loss of life	14- 33	–	–
Reduced agricultural productivity	0-170	100	33
Total (rounded)	**440-1,300**	**770**	**250**

* These costs do not take into account the billions of dollars that have been spent to avoid flood damages by constructing flood-control dams, channelization projects, drainage projects, sediment traps, and other flood-mitigation measures. Some portion of the cost of building and maintaining these facilities could also reasonably be allocated to controlling the sediment-associated flood damages.

Source: Original Conservation Foundation research.

million ($8.0 million in 1980 dollars).[93]* As chapter 3 indicated, this total represents 1.4 percent of the total amount of erosion in Illinois.[95] Assuming that nationally 0.75 to 1.5 percent of gross erosion is deposited in and must be removed from ditches, and that the average removal cost is the same as in Illinois, the national annual cost of ditch cleaning would be $90 to $180 million (figure 5.13).

Irrigation Canals

In 1978, a total of $281 million ($330 million in 1980 dollars) was spent to operate and maintain U.S. irrigation canals, according to the 1978 Census of Agriculture.[96] A significant portion of this money was used to finance the removal of sediment and weeds from the channels.

One researcher estimated that, in 1940, $10 million ($61 million in 1980 dollars) was spent to deal with problems caused by sedimentation; that amount represented approximately one-fourth of the total annual operation and maintenance costs for irrigation enterprises.[97]

*A similar study in Ohio concluded that the sediment-removal costs in the areas studied were about $1 million per year in 1979 dollars.[94]

Figure 5.13
Costs to Water-Conveyance Facilities
(million dollars)

Type of impact	Range of estimates	Single-value estimate	Cropland's share
Drainage ditches			
Flooding damages		no estimate	
Sediment removal	90-180	120	50
Irrigation canals			
Sediment removal	30- 70	50	25
Weed control	15- 50	35	25
Pumping water			
Increased pumping costs		less than $1 million	
Equipment maintenance costs		no estimate	
Total (rounded)	**140-300**	**200**	**100**

Source: Original Conservation Foundation research.

Another estimated in 1966 that removing sediment from both irrigation canals and drainage ditches that year cost $34 million ($79 million in 1980 dollars).[98]

Apparently no one has estimated the impacts of sediment and nutrients on weed control, but this can also be a large maintenance cost for many canal systems. For instance, the small Kennewick Canal District near the mouth of the Yakima River in Washington State spends about $60,000 a year on herbicides to control weed growth in its 44 miles of canals. However, because almost the entire water supply for this district is composed of irrigation return flow with high nutrient levels, those expenditures are probably above average.[99]

The operations chief of the U.S. Bureau of Reclamation has estimated that 10 to 20 percent of the bureau's maintenance costs are spent on removing sediment and 5 to 15 percent on removing weeds from canals.[100] Combining these estimates with the Census of Agriculture's results indicates that each year $30 to $70 million may be spent on sediment removal and $15 to $50 million on weed control.*

*Irrigation canals provide an interesting example of the divergence between sediment damages and the benefits that would result from sediment control. Approximately 95,000 miles of the nation's 110,000 miles of irrigation canals are unlined.[101] Many of these unlined canals depend on sediment suspended in intake water to prevent erosion of their canal banks. When this sediment is removed, serious and costly erosion can occur in the canal.[102]

Most of these costs can be attributed to sediment and nutrients from agricultural lands, because nutrient sources (municipal treatment plants and industries) are more likely to be located downstream of canal diversions.

Pumping Costs

Sediment may increase water-pumping costs by requiring extra electricity for the particles to be pumped along with the water. However, if (as chapter 3 estimated) suspended sediment annually requires 5 or 6 million additional kilowatt hours of electricity to pump the water, the extra cost of this electricity would be less than $1 million a year.

But suspended sediment may further increase pumping costs by necessitating increased maintenance expenditures, although there seems to be no estimate of how large these expenditures might be. Nor are there any data on how sediment affects pumping efficiency. Nevertheless, these costs could be much larger than the increased pumping costs. A 1 percent reduction in average pumping efficiency would increase annual pumping costs by $4 to $5 million per year.[103] However, because there is no information on these effects, no estimate of their magnitude is made in this report.

Water Treatment

Sediment and its associated contaminants can substantially increase the problems and, therefore, the costs of providing safe drinking water and process water for off-stream users. Especially turbid surface-water supplies that require the use of large settling basins and chemical coagulants as well as filters may increase the cost of water treatment up to 35 percent, depending on the size of the plant.[104]* Removing nitrogen or pesticides can more than double the cost.

Over the years, several attempts have been made to determine the annual cost to the United States of treating public water supplies to remove sediment and associated contaminants. A 1947 estimate placed the "cost of water purification as a result of excess turbidity" at $5 million annually ($18 million in 1980 dollars).[105] A 1966 estimate put the annual cost for "removing the excess turbidity from public water supplies" at $14 million ($39 million in 1980 dollars) and for

*The costs, however, vary in a complicated manner with the amount and type of sediment suspended in the raw water supply. To force proper flocculation, a settling basin may actually have to be larger and more chemicals may have to be added at low suspended sediment levels. With higher sediment levels, the sediment can speed the flocculation and sedimentation process.

"miscellaneous sediment removal, cleaning and adding maintenance" at $31 million ($87 million in 1980 dollars).[106] One study noted that additional coagulants, disinfectants, and filtration systems are among the major expenses in the operation of many water-treatment systems.[107] The U.S. Senate Committee on Agriculture and Forestry estimated in 1974 that $25 million ($39 million in 1980 dollars) was being spent on "added maintenance and turbidity removal by industry and cities."[108] One analyst estimated that the savings to municipal water treatment resulting from abatement of conventional water pollutants from all sources would amount to $0.6 to $1.2 billion in 1978 dollars ($0.7 to $1.4 billion in 1980 dollars).[109]

A 1979 benefit assessment of the water-quality plan for Kansas analyzed cost savings for water clarification, softening, and sludge handling for the state's different river basins. The report concluded that the total 1980 cost savings for Kansas alone would be almost $32 million in 1974 dollars (about $50 million in 1980 dollars) for "full treatment" of contaminants contributed by nonpoint sources.[110] This estimate, however, seems a bit high.* A report prepared by the Baltimore city government estimated that nonpoint-source-caused eutrophication in that city's Loch Raven reservoir adds between $200,000 and $250,000 to Baltimore's water-treatment costs each year.[111] An analysis of possible off-farm costs in a region of Michigan concluded that the cost savings in three water-treatment plants there are unlikely to exceed $40,000.[112]

An 1979 EPA analysis of water-treatment costs between November 1976 and January 1979 indicated that using conventional treatment (including sedimentation) to removed suspended sediment and associated contaminants from municipal water supplies cost new plants 2 to 8 cents per 1,000 gallons more than did using filtration alone.[113]† If it is assumed that the current national average (including old and new plants) falls between 0.5 and 5 cents per 1,000 gallons, the annual cost of increased water-treatment expenditures can be estimated

---

*The estimate is equivalent to $358,000 annually per million gallons per day of water treated. Multiplying this cost by the total amount of water withdrawn for public water supplies gives a national annual cost of about $7.9 billion per year.

†The average cost of supplying drinking water from a newly constructed direct-filtration water-treatment plant in 1978 ranged from $0.63 per 1,000 gallons for a plant that treated 1 million gallons per day (suitable for a population of 3,500 to 4,500) down to $0.122 per 1,000 gallons for a plant that treated 100 million gallons per day (75 cents and 15 cents, respectively, in 1980 dollars).[114]

at $50 to $500 million.* The single-value estimate can be assumed to be $100 million, of which cropland's share is $30 million.

Municipal and Industrial Use

Although water-treatment facilities deal effectively with contaminants that can cause illnesses, they do not remove many dissolved minerals and salts that can interfere with the efficient operation and durability of water-using equipment in industries and homes (see chapter 4).

High concentrations of dissolved solids can be a particularly serious problem for industrial water users. The salts are incompatible with many physical and chemical processes for which water is needed. They also induce scaling and corrosion of machinery. As a result, most water supplies must be demineralized before they are used in industrial boilers. In addition, organic particles and algae associated with runoff can interfere with this demineralization and must be removed from the water supply before it can be treated for salts.[115] Treatment processes to remove dissolved solids are expensive.[116] In 1975, demineralization of boiler feed-water cost between $2.14 and $3.13 per 1,000 gallons (between $3.04 and $4.44 in 1980 dollars).[117]†

A 1973 study estimated that the overall annual cost of hardness and total dissolved solids (TDS), both of which are related to water salinity, was $0.65 to $3.45 billion in 1970 dollars, with a single-value estimate of $1.75 billion ($1.3, $6.7, and $3.4 billion, respectively, in 1980 dollars).[118] Over one-third of those costs were experienced by households using treated surface water.[119] A more recent study of the costs of salinity in the Colorado River indicated similarly large costs. Damages to municipal and industrial users obtaining their water supply from that river were estimated at about $80 million per year in 1982 dollars ($69 million in 1980 dollars).[120]

In most parts of the United States, much of the water hardness and TDS comes from natural sources. And groundwater typically causes greater damage than surface water because of its naturally higher hardness.[121] Some of the salts, however, are contributed by runoff from agricultural lands.

---

*These calculations assume that 5 million gallons per day of industrial water are treated to drinking water standards.

†The 1975 price deflator was used to convert these dollar figures to their 1980 equivalents.

Figure 5.14
Other Off-Stream Effects
(million 1980 dollars)

Type of impact	Range of estimates	Single-value estimate	Cropland's share
Municipal and industrial users	500-1,300	900	300
Steam electric power plants			
Maintenance of cooling systems	17- 28	20	7
Increased efficiency*	(−)100-(−)400	(−)150	(−)50
Irrigation			
Sealing soil surface		no estimate	
Salinity	3- 34	25	15
Value of nutrients*	(−)15-(−)37	(−)25	(−)8
Total (rounded)	**400- 920**	**770**	**260**

* Effects that generate cost savings are entered as negative numbers in this table of costs.

Source: Original Conservation Foundation research.

Several national studies have attempted to sort out the amount of damages resulting from natural and human sources of hardness and dissolved solids. The most recent of these concludes that achieving the goals of the Clean Water Act will generate $0.5 to $1.3 billion in benefits for households and industries in 1978 dollars ($0.6 to $1.5 billion in 1980 dollars). Nonpoint sources are estimated to account for 80 to 85 percent of TDS loadings. Assuming that they are responsible for the same percentage of damages would give a range of $0.5 to $1.3 billion worth of damages, with a single-value estimate of $0.9 billion (figure 5.14).

In addition, although steam electric power plants and other water-cooling facilities do not require high-quality water, sediment and algae can interfere with the efficient operation of these facilities. This can lead to system breakdowns and financial losses. For instance, each time Pickering, Ontario, had to close its power plant because algae was clogging water lines (see chapter 4), the town lost $40,000 in revenues.[122] The heat exchangers that the utility was eventually forced to install cost $2.7 million in 1975 dollars ($3.8 million in 1980 dollars).[123]

Another study concluded that full treatment of point and nonpoint sources of pollution in Kansas would result in benefits to the state's steam electric power generation totaling $6.4 million in 1975 dollars

($9 million in 1980 dollars).[124] Again, this estimate seems high.*

An analysis by the Electric Power Research Institute estimated that 58.7 percent of U.S. steam-electric generating plants use chlorination for biofouling control and that the cost of removing algae from condensers by chlorination amounts to $62 to $105 per installed megawatt of generating capacity per year.[127]† In 1980, there were approximately 540,000 megawatts of thermoelectric installed capacity, essentially all of which was cooled by surface water.[128] Thus, the total cost of controlling algae would have been between $20 and $33 million. Because approximately 15 percent of nutrients in surface water come from point sources, the total cost attributable to nonpoint sources would have been between $17 and $28 million. A single-value estimate would be $20 million, with cropland's share being $7 million.

At least partially offsetting these costs is the possible increased generating efficiency that could result from turbid ponds being cooler than clear ponds. The average efficiency of steam electric power plants is currently about 33 percent.[129] Using basic laws of thermodynamics, a two-degree-centigrade reduction in the temperature of the cooling water could increase the efficiency by as much as 0.4 percent.[130] Assuming that the actual increase in efficiency could range from a national average of 0.1 percent to 0.4 percent gives a range of benefits of $100 million to $400 million per year, with a single-value estimate of a $150 million benefit (figure 5.14).[131]‡ Cropland's share of the benefit would be $50 million.

---

*The Kansas estimate is equivalent to $86 worth of cost savings for each million gallons of water withdrawn. Nationally, 150 billion gallons of water a day are withdrawn from fresh surface water by steam electric power plants for cooling purposes.[125] At $86 per million gallons, the total cost savings would be $4.7 billion per year, equivalent to almost 15 percent of the total 1980 production costs for steam electric power plants using fresh surface water for cooling.[126]

†If these expenditures are not made, the costs to a generating facility can increase significantly. The plant may also experience costs associated with sediment in the cooling water, but there is apparently no information available upon the prevelance of such problems. As a result, no such costs were estimated for this report.

‡The single-value estimate is at the low end of this range because: (*a*) the difference in temperatures is only noticeable for part of each year; (*b*) the data on temperature differences pertained to waters that were much more turbid than would normally be used for cooling; and (*c*) much cooling water is recycled, which normally removes turbidity from the water. (In fact, if the turbidity were not removed, it could reduce the rate at which heat was lost from the cooling pond, thereby increasing rather than decreasing the temperature of the cooling water.)

Irrigation

Farmers using irrigation water containing sediment and other erosion-related contaminants may experience increased costs in addition to those caused by sedimentation in irrigation canals. For instance, fine silt may form a crust on the surface of the soil, lowering crop yields by reducing infiltration and seed germination. In addition, dissolved salts may lower crop yields. There is apparently no information available that would allow the magnitude of the fine silt problem to be estimated, but it is possible to make a rough estimate of salt damage.[132]

Various studies completed in the early 1970s on the damages to agriculture from saline irrigation water concluded that the total damages ranged from $530 to $960 million for salinity contributed by human activities in irrigation supplies west of the Mississippi River.[133] More recent studies of the costs of salinity in the lower Colorado River basin suggest much lower costs.[134] Direct agricultural damages there are estimated to be $1.7 million a year.[135] But much of the salt comes from natural sources, and agricultural damages attributable to agricultural sources amount to about $1.5 million.[136] That $1.5 million is spread over approximately one million acres of land, giving an average damage cost of about $1.50 an acre. It is extremely unlikely that the average national damage costs are higher than those found in the Colorado basin. Therefore, assuming that damages nationally are between $0.10 and $1.00 per irrigated acre,[137] the net annual cost of salinity is between $3 and $34 million dollars, with a single-value estimate of $25 million (figure 5.14). Cropland's share of this cost is $15 million.

However, contaminated irrigation water also can benefit farmers if it contains nutrients. For instance, each acre-foot of water containing 1.0 and 0.5 milligrams per liter of nitrogen and phosphorus, respectively, would contribute the equivalent of $1.50 worth of fertilizer at 1980 prices.[138]

Irrigation in the United States consumes one million acre-feet of water from surface-water sources annually.[139] In 1975, according to water-quality monitoring by the U.S. Geological Survey, surface-water supplies contained an average of 0.85 milligrams per liter of nitrogen (with most supplies falling in the range of 0 to 2.75 milligrams per liter) and a mean of 0.24 milligrams per liter of phosphorus (with most supplies falling in the range of 0 to 0.75 milligrams per liter).[140] Agricultural sources are commonly estimated to be responsible for 80 to 90 percent of total pollutant loadings of both nutrients.[141] Assuming that agriculture annually supplies an average of 0.5 to 1.0 milligrams per liter of nitrogen and 0.10 to 0.30 milligrams per liter of phosphorus to irrigation water supplies, the net benefit of agriculture-supplied nutrients in irrigation water would be between $15 and $37 million, with a single-value benefit of $25 million (figure 5.14). Cropland's share of the benefit would be $8 million.

CONCLUSIONS

Combining all the cost estimates made in this chapter indicates that erosion-related pollutants impose net damage costs of $3.2 to $13 billion per year in the United States, with a single-value estimate of $6.1 billion (figure 5.15). Cropland's share would be $2.2 billion. Many of these costs—for instance, salinity damages to household and industrial water-using equipment—are only remotely related to erosion and could continue even if erosion were stopped completely. The amount of damage directly related to sediment would be approximately $3.2 to $3.7 billion for all sources and $1.1 to $1.3 billion for cropland erosion,* with the remainder due to sediment-associated contaminants.

Soil erosion, it is clear, is imposing substantial costs on the United States at the present time. Attempts to control it efficiently, if successful, should pay substantial dividends.

---

*These estimates arbitrarily assume that sediment alone is responsible for one-third to one-half of the damages to recreation and other instream uses—when the nutrients and sediments actually act together to cause the problem. In addition, since some potentially very significant impacts of soil erosion—both positive and negative—have not been estimated in this chapter, the full costs could differ substantially from these overall estimates.

Figure 5.15
Summary of Damage Costs
(million 1980 dollars)

Type of impact	Range of estimates	Single-value estimate	Cropland's share
In-stream effects			
Biological impacts		no estimate	
Recreational	950-5,600	2,000	830
Water-storage facilities	310-1,600	690	220
Navigation	420- 800	560	180
Other in-stream uses	460-2,500	900	320
Subtotal—In-stream (rounded)	2,100-10,000	4,200	1,600
Off-stream effects			
Flood damages	440-1,300	770	250
Water-conveyance facilities	140- 300	200	100
Water-treatment facilities	50- 500	100	30
Other off-stream uses	400- 920	800	280
Subtotal—Off-stream (rounded)	1,100-3,100	1,900	660
Total—all effects (rounded)	**3,200-13,000**	**6,100**	**2,200**

Source: Original Conservation Foundation research.

REFERENCES

1. Carl B. Brown, "Perspective on Sedimentation—Purpose of Conference" (Proceedings of the First Federal Interagency Sedimentation Conference, May 6-8, 1947), p. 6.

2. Kim S. Harris and Wesley D. Seitz, "Sediment Damages in Water," University of Illinois at Urbana-Champaign, n.d., p. 10. Agricultural land acreage (the measure of acreage used in the study), including cropland, rangeland, and pastureland, amounted to 1,058 million acres in 1978. From U.S. Department of Agriculture, as quoted in The Conservation Foundation, *State of the Environment: An Assessment at Mid-Decade* (Washington, D.C.: The Conservation Foundation, 1984), figure 3.10, p. 147.

3. M. T. Lee et al., "Economic Analysis of Erosion and Sedimentation: Hambaugh-Martin Watershed," prepared for University of Illinois at Urbana-Champaign, Illinois Agricultural Experiment Station, and State of Illinois, Institute for Environmental Quality, July 1974, pp. 27, 31.

4. Edwin H. Clark II, "Estimated Effects of Non-point Source Pollution" (Washington, D.C.: The Conservation Foundation, 1984), table 6.

5. Judd Hammack and Gardner Mallard Brown, Jr., *Waterfowl and Wetlands: Toward Bioeconomic Analysis*, prepared for Resources for the Future (Baltimore: Johns Hopkins University Press, 1974); Robert Prescott-Allen and Christine

Prescott-Allen, *What's Wildlife Worth?* (London: Earthscan, 1982); and Norman Myers, *The Sinking Ark* (Oxford: Pergamon Press, 1979).

6. Sterling Hobe Corporation, "Regulatory Impact Analyses on Regulations Affecting Hydrocarbon Operations within the Channel Islands and Point Reyes-Farallon Islands National Marine Sanctuaries," appendices, vol. 1, prepared for U.S. Department of Commerce, Contract no. NA-81-SAC-00724, January 1982, p. B-93.

7. Rules of State of Florida, Department of Pollution Control, chapters 17-ll, Assessment of Damages; adopted May 15, 1971.

8. U.S. Department of Commerce, National Oceanic and Atmospheric Administration, *The Use of Economic Analysis in Valuing Natural Resource Damages* (Washington, D.C.: U.S. Government Printing Office, 1984).

9. Outdoor Recreation Policy Review Group, *Outdoor Recreation for America 1983: An Assessment Twenty Years After the Report of the Outdoor Recreation Resources Review Commission* (Washington, D.C.: Resources for the Future, Inc., 1983), pp. 15-16.

10. Ibid.

11. Ibid.

12. National Research Council, Committee on Impacts of Emerging Agricultural Trends on Fish and Wildlife Habitat, *Impacts of Emerging Agricultural Trends on Fish and Wildlife Habitat* (Washington, D.C.: National Academy Press, 1982), p. 199.

13. C. Tim Osborn and Robert N. Shulstad, "Controlling Agricultural Soil Loss in Arkansas' North Lake Chicot Watershed: An Analysis of Benefits," *Journal of Soil and Water Conservation* 38, no. 6 (1983):509-12.

14. Alfred Birch, Carmen Sandretto, and Lawrence W. Libby, *Toward Measurement of the Off-Site Benefits of Soil Conservation*, Agricultural Economics Report no. 431 (East Lansing, Mich.: Michigan State University, Department of Agricultural Economics, 1983), p. 41.

15. JACA Corporation, *An Assessment of Economic Benefits of 28 Projects in the Section 314 Clean Lakes Program*, prepared for U.S. Environmental Protection Agency, Office of Water Planning and Standards (Fort Washington, Pa.: JACA Corporation, 1980), p. 12.

16. Rodney W. Olson and Carl L. Armour, "Economic Considerations for Improved Livestock Management Approaches for Fish and Wildlife in Riparian/Stream Areas," in *Proceedings of the Forum—Grazing and Riparian/Stream Ecosystems*, held at Denver, Colo., November 3-4, 1978 (*Trout Unlimited*, 1979), p. 70. Value estimates were based on calculations in W.E. Martin, J.C. Tinney, and R.L. Gum, *A Welfare Economic Analysis of the Potential Competition between Hunting and Cattle Ranching,* Arizona Agricultural Experiment Station Paper no. 238, 1978.

17. William J. Vaughan and Clifford S. Russell, *Freshwater Recreational Fishing—The National Benefits of Water Pollution Control*, prepared for Resources for the Future (Baltimore, Md.: Johns Hopkins University Press, 1982).

18. Ibid., p. 157.

19. Ibid., p. 160.

20. Ibid., p. 160.

21. Ibid., p. 49.

22. Ibid., pp. 161-62.

23. A. Myrick Freeman III, *Air and Water Pollution Control-A Benefit-Cost Assessment* (New York: John Wiley and Sons, 1982).

24. Ibid., p. 161.

25. According to the model used in the RFF study, cropland erosion is responsible for 37 to 38 percent of the total BOD loadings from nonpoint sources. L. P. Gianessi and H. M. Peskin, "Analysis of National Water Pollution Control Policies: 2. Agricultural Sediment Control," *Water Resources Research* 17, no. 4 (1981):804.

26. *Inside EPA*, January 6, 1984, pp. 10-13.

27. R. E. Sparks, *Effects of Sediment on Aquatic Life* (Havana, Ill.: Illinois Natural History Survey, 1977), p. 19.

28. Brown, "Perspective on Sedimentation," p. 6; and U.S. Department of Commerce, Bureau of Reclamation, *Construction Review,* table E-1, December 1978 and more recent issues.

29. U.S. Department of Commerce, *Construction Review*, December 1978 and more recent issues.

30. Harris and Seitz, "Sediment Damages in Water," table 1.

31. U.S. Department of Commerce, *Construction Review*, December 1978 and more recent issues.

32. Vito A. Vanoni, ed., *Sedimentation Engineering*, prepared by the American Society of Civil Engineers Task Committee for the preparation of the Manual on Sedimentation of the Sedimentation Committee of the Hydraulics Division (New York: American Society of Civil Engineers, 1977), p. 617.

33. Ven Te Chow, ed., *Handbook of Applied Hydrology: A Compendium of Water-Resources Technology* (New York: McGraw-Hill Book Co., 1964), pp. 17-29.

34. R. P. Beasley, *Erosion and Sediment Pollution Control*, (Ames, Iowa: Iowa State University Press, 1972), p. 18.

35. U.S. Army Corps of Engineers, Tulsa District, "Upper Little Arkansas River Watershed Survey Report and Environmental Impact Statement," vol. 1, October 1981, pp. 8-37 to 8-43.

36. Personal communication from Owen Lee, U.S. Soil Conservation Service, May 13, 1983.

37. Capacity for withdrawal purposes was estimated to increase only 13 million acre-feet (or 8 percent) between 1970 and 1980 (Walter B. Langbein, "Dams, Reservoirs and Withdrawals for Water Supply—Historic Trends," prepared for U.S. Department of the Interior, Geological Survey, Open-File Report no. 82-256, [Washington, D.C.: U.S. Department of the Interior, 1982], p. 6). However, most of the capacity is provided for nonwithdrawal purposes. Total capacity appears to have increased approximately 50 million acre-feet during the same period. The U.S. Department of Agriculture is constructing about 200,000 acre-feet of capacity a year in its PL 566 program (personal communication from Herman Calhoun, March 28, 1985), but this is well within the probable error range of the above estimate.

38. Chow, *Handbook of Applied Hydrology*, pp. 17-30.

39. L. C. Gottschalk, "Reservoir Sedimentation," in Chow, *Handbook of Applied Hydrology*, pp. 17-30.

40. U.S. Department of the Interior, Geological Survey, Office of Water Data Coordination, *Notes on Sedimentation Activities, Calendar Year 1980* (Washington, D.C.: U.S. Government Printing Office, 1981), p. 4.

41. Vanoni, *Sedimentation Engineering*, p. 617.

42. A. G. Taylor et al., "Costs of Sediment in Illinois Roadside Ditches and Rights-of-Way," Collection of Reports on Effects and Costs of Erosion in Illinois, Illinois Environmental Protection Agency, October 1978, p. 8.

43. A. G. Taylor et al., "Estimated Annual Costs of Sedimentation in Illinois Lakes and Impoundments," Collection of Reports on Effects and Costs of Erosion in Illinois, Illinois Environmental Protection Agency, October 1978, p. 5.

44. S. A. Peterson, "Dredging and Lake Restoration," in *Lake Restoration* (Proceedings of a national conference, held at Minneapolis, August 22-24, 1978), EPA 440/5-79-001 (Washington, D.C.: U.S. Government Printing Office, 1979), p. 113.

45. U.S. Comptroller General, "Dredging America's Waterways and Harbors: More Information Needed on Environmental and Economic Issues," Report to Congress, CED-77-74 (Washington, D.C.: U.S. General Accounting Office, June 28, 1977), p. 19.

46. Peterson, "Dredging and Lake Restoration," p. 113.

47. A. G. Taylor, "Summary of the Agriculture Task Force Water Quality Plan Recommendations," prepared by Illinois Institute for Environmental Quality, Task Force on Agriculture Non-point Sources of Pollution, for Illinois Environmental Protection Agency, March 1978, p. 7.

48. U.S. Comptroller General, "Dredging America's Waterways and Harbors," pp. 46-52.

49. U.S. Army Corps of Engineers, *Deep-draft Access to the Ports of New Orleans and Baton Rouge, Louisiana—Feasibility Study*, vol. 1, *Main Report and Final Environmental Impact Statement* (New Orleans District, La.: U.S. Army Corps of Engineers, 1981), p. EIS-47.

50. Taylor et al., "Estimated Annual Costs of Sedimentation in Illinois Lakes and Impoundments," p. 5.

51. San Francisco Bay Area Water Quality Planning Program, "Water Quality Management Plan—Erosion-Related Water Quality Problems," Technical Memorandum no. 55, May 1980, revised September 1980, p. 7.

52. Peterson, "Dredging and Lake Restoration," p. 113.

53. If the capacity were not replaced for an average of 20 years, the costs should be discounted by about 85 percent at a 10 percent discount rate and 60 percent at a 5 percent discount rate. If the capacity were to be replaced in an average of 10 years, the costs should be discounted by about 60 percent with a 10 percent annual discount rate, and 40 percent with a 5 percent annual discount rate.

54. Robert Stavins, *Trading Conservation Investments for Water* (Berkeley, Calif.: Environmental Defense Fund, Inc., March 1983), p. 177.

55. U.S. Department of the Interior, Bureau of Reclamation, *Geary Project—Oklahoma Concluding Report*, table 22—"Construction and OM&R Costs, Jan. 1980," p. 72 (Washington, D.C.: U.S. Department of the Interior, 1981).

56. A. R. Castorina, "Reservoir Improvements as a Key to Source Quality Control," *Journal of American Water Works Association* 72, no. 1 (1980):28.

57. U.S. Environmental Protection Agency, Office of Water Regulations and Standards, *Restoration of Lakes and Inland Waters*, EPA 440/5-81-010 (Washington, D.C.: U.S. Environmental Protection Agency, 1980), pp. 89-92.

58. Ibid., p. 115; and E. B. Welch et al., "Long-term Lake Recovery Related to Available Phosphorus"; cited in William P. Stack and Jane C. Gottfredson, "Data Evaluation for Determination of Eutrophication Control Criteria: Loch Raven Reservoir Project," prepared for City of Baltimore, Water Quality Management Office, December 1980, p. 32.

59. Justine Welch, *The Impact of Inorganic Phosphates in the Environment*, prepared for U.S. Environmental Protection Agency, Office of Toxic Substances, EPA-560/1-78-003 (Washington, D.C.: U.S. Environmental Protection Agency, 1978), p. 47.

60. Cayuga County Environmental Management Council, "A Plan to Control Nonpoint Pollution in the Dutch Hollow Brook Watershed," vol. 3, "Appendix," Auburn, N.Y., November 1980, pp. 20-21.

61. Federal Water Pollution Control Act, P.L. 92 500, 314.

62. Armstrong, "Phosphorus Transport across the Sediment-Water Interface," p. 169-70.

63. Personal communication with Frank LaPensee, Office of Water, U.S. Environmental Protection Agency, October 7, 1984.

64. State of Wisconsin, Legislative Audit Bureau, "The Inland Lake Renewal Program," Audit no. 81-26, 1981.

65. EPA has spent about $10 million a year during the program's active years. In Wisconsin, state and local governments spent almost $9 for every EPA dollar.

66. U.S. Coast Guard, "Statistics of Casualties: Fiscal Year 1980," *Proceedings of the Marine Safety Council*, February 1983.

67. U.S. Army Corps of Engineers, *Water Spectrum* 6 (Winter 1974-75):4.

68. Birch, Sandretto, and Libby, *Toward Measurement of the Off-Site Benefits of Soil Conservation*, pp. 47-48.

69. Personal communication with Charles Hummer, Jr., U.S. Army Corps of Engineers, September 7, 1984; and "Summary of Dredging Activities Corps and Industry Dollars and Yard (millions)" (chart) (Fort Belvoir, Va.: U.S. Department of the Army, 1982).

70. Personal communication with Hummer.

71. M. B. Boyd et al., "Disposal of Dredge Spoil: Problem Identification and Assessment and Research Program Development," U.S. Army Engineer Waterway Experiment Station, Vicksburg, Miss., Technical Report no. H-72-8, 1972, p. 3.

72. Freeman, *Air and Water Pollution Control*, p. 170.

73. *Staff Report to the National Commission on Water Quality* (Washington, D.C.: U.S. Government Printing Office, 1976), p. III-286.

74. Freeman, *Air and Water Pollution Control*, pp. 163-64.

75. Welch, *The Impact of Inorganic Phosphates in the Environment,* p. 51.

76. Marc Ribaudo, C. Edwin Young, and Donald Epp, *Recreational Benefits from an Improvement in Water Quality at St. Albans Bay, Vt.*, Staff Report no. AGES840127, March 1984.

77. Douglas A. Greenley, Richard G. Walsh, and Robert A. Young, *Economic Benefits of Improved Water Quality: Public Perceptions of Option and Preservation Values*, Studies in Water Policy and Management no. 3 (Boulder, Colo.: Westview Press, 1982), p. 117.

78. Ann Fisher and Robert Raucher, "Intrinsic Benefits of Improved Water Quality: Conceptual and Empirical Perspectives," in V. Kerry Smith, ed., *Advances in Applied Microeconomics* (Greenwich, Conn.: JAI Press, forthcoming); cited in Environmental Defense Fund, "The Tuolumne River: Preservation or Development? An Economic Assessment—Summary Report" (Berkeley, Calif.: Environmental Defense Fund, 1983).

79. Freeman, *Air and Water Pollution Control*, p. 170.

80. U.S. Water Resources Council, *The Nation's Water Resources 1975-2000,* vol. 3, *Analytical Data Appendix 1, Social, Economic, and Environmental Data*

(Washington, D.C.: U.S. Government Printing Office, 1978), pp. 80-81. The 1980 estimates were developed by averaging the 1975 and 1985 estimates produced by the Water Resources Council, and adjusting from 1967 to 1980 dollars using the GNP deflators.

81. Beasley, *Erosion and Sediment Pollution Control*, p. 20.

82. U.S. Water Resources Council, *The Nation's Water Resources 1975-2000,* vol. 3, *Analytical Data Appendix 1,* pp. 80-81.

83. Brown, "Perspective on Sedimentation," p. 7.

84. E. C. Ford, "Upstream Flood Damage," *Journal of Soil and Water Conservation* 19, no. 6 (1964).

85. Vanoni, *Sedimentation Engineering*, p. 616.

86. State of North Carolina, U.S. Water Resources Council, and State of South Carolina, "Yadkin-Pee Dee River Basin: Level B Comprehensive Water Resources Study—Recommended Plan," August 1981, p. 68.

87. John C. Briggs and John F. Ficke, *Quality of Rivers of the United States, 1975 Water Year, Based on the National Stream Quality Accounting Network (NASQAN)*, prepared for U.S. Department of the Interior, Geological Survey, Open Report no. 78-200 (Reston, Va.: U.S. Geological Survey, 1977), p. 42, gives maximum sediment concentrations; U.S. Water Resources Council, *The Nation's Water Resources 1975-2000,* vol. 3, *Analytical Data Appendix 1,* pp. 80-81, says flood damages in 1980 were assumed to be the average of 1975 and 1985 damages, after adjusting by the GNP deflator.

88. Brown, "Perspective on Sedimentation," p. 7.

89. Ford, "Upstream Flood Damage," p. 232.

90. The Conservation Foundation, *State of the Environment 1982* (Washington, D.C.: The Conservation Foundation, 1982), p. 105.

91. U.S. Department of Agriculture, *1980 Appraisal Part II: Soil, Water, and Related Resources in the United States—Analysis of Resource Trends* (Washington, D.C.: U.S. Government Printing Office, 1981), p. 149.

92. U.S. Department of Commerce, *Statistical Abstracts of the United States: 1984*, 104th ed. (Washington, D.C.: U.S. Government Printing Office, 1984), table 1162, p. 661. Gross farm income from crops was $72.7 billion dollars in 1980.

93. Taylor et al., "Costs of Sediment in Illinois Roadside Ditches and Rights-of-Way," p. 2. Using implicit GNP deflator for state and local government purchase of goods and services.

94. Harris and Seitz, "Sediment Damages in Water," p. 3 and table I.

95. Illinois Institute for Environmental Quality, Task Force on Agriculture Nonpoint Sources of Pollution, "Final Report," December 1978, p. 18.

96. U.S. Department of Commerce, Bureau of the Census, *1978 Census of Agriculture*, vol. 4, Irrigation (Washington, D.C.: U.S. Government Printing Office, 1982), tables 17 and 20, p. 278.

97. Brown, "Perspective on Sedimentation," p. 6.

98. A. R. Robinson, "Sediment: Our Greatest Pollutant?", *Agricultural Engineering*, August 1971, p. 406.

99. Personal communication from Paul Chasco, Kennewick Irrigation District, August 25, 1983.

100. Letter from Jerome Schaack, chief, Water Operations and Maintenance Branch, Bureau of Reclamation, U.S. Department of the Interior, to Edwin H. Clark II, August 22, 1984.

101. U.S. Department of Commerce, Bureau of the Census, *1978 Census of Agriculture*, vol. 4.

102. George V. Sabol, "Desilted Irrigation Water: A Case Study," in *Irrigation and Drainage in the Nineteen-Eighties* (New York: American Society of Civil Engineers, 1979), pp. 272-74.

103. It makes the same assumptions in amount of water pumped as other estimates.

104. U.S. Environmental Protection Agency, Municipal and Environmental Research Laborary, *Estimating Water Treatment Costs*, vol. 1, *Summary,* EPA 600/2-79-162a (Cincinnati: U.S. Environmental Protection Agency, 1979), pp. 54-67.

105. Brown, "Perspective on Sedimentation," p. 7, using GNP deflator for state and local government purchases of goods and services.

106. Robinson, "Sediment: Our Greatest Pollutant?" p. 406, using GNP deflator for state and local government purchases of goods and services.

107. North Carolina State University, Biological and Agricultural Engineering Department, North Carolina Agricultural Extension Service, "Best Management Practices for Agricultural Nonpoint Source Control: III. Sediment, Raleigh, N.C., n.d., p. 2.

108. U.S. Senate, Committee on Agriculture and Forestry, *Conservation of the Land and the Use of Waste Materials for Man's Benefits*, Committee Print, 94th Cong., 1st sess., March 25, 1975, p. 14.

109. Freeman, *Air and Water Pollution Control*, pp. 168-70.

110. M. Jarvin Emerson and Hossein Akhavipour, *Benefit Assessment of Kansas Water Quality Management Plan* (Manhattan, Kans.: Kansas State University, 1979), p. 123, using GNP deflator for state and local government purchase of goods and services.

111. William P. Stack and Jane C. Gottfredson, *Options for a Stepwise Eutrophication Control Strategy: Loch Raven Reservoir Project* (Baltimore: Water Quality Management Office, 1981), pp. 90-93.

112. Birch, Sandretto, and Libby, *Toward Measurement of the Off-Site Benefits of Soil Conservation*, p. 46.

113. U.S. Environmental Protection Agency, *Estimating Water Treatment Costs*, vol. 1, *Summary*, pp. 62-66, using GNP deflator for state and local government purchase of goods and services.

114. Wayne B. Solley, Edith B. Chase, and William B. Mann IV, *Estimated Use of Water in the United States in 1980*, U.S. Department of the Interior, Geological Survey, Circular no. 1001 (Reston, Va.: U.S. Geological Survey, 1983), p. 10.

115. R. D. Taylor, J. E. Dailey, and G. E. Rohlich, "Wastewater Effluent Discharge to Cooling Lakes," *Journal of Environmental Engineering* 103 (1977); cited in Welch, *The Impact of Inorganic Phosphates in the Environment*, p. 50.

116. Panhandle Regional Planning Commission, *Plan Summary Report for the Canadian Basin Water Quality Management Plan*, prepared for Texas Department of Water Resources, August 1978, revised June 1981, p. II-D-11.

117. Taylor, Dailey, and Rohlich, "Wastewater Effluent Discharge to Cooling Lakes"; cited in Welch, *The Impact of Inorganic Phosphates in the Environment*, p. 50.

118. Dennis P. Tihansky, "Economic Damages from Residential Use of Mineralized Water Supply," *Water Resources Research* 10, no. 2 (1974):145.

119. Ibid., p. 153.

120. Estimated from U.S. Department of the Interior, Bureau of Reclamation, Colorado River Water Quality, *Colorado River Water Quality Improvement Program, Status Report* (Washington, D.C.: U.S. Government Printing Office, 1983), pp. 5, 21. Total damages are $113 million, and municipal and industrial damages account for 70 percent of these.

121. Tihansky, "Economic Damages from Residential Use of Mineralized Water Supply," pp. 151-52.

122. Welch, *The Impact of Inorganic Phosphates in the Environment*, p. 49.

123. Ibid.

124. Emerson and Akhavipour, *Benefit Assessment of Kansas Water Quality Management Plan*, p. 152.

125. Solley, Chase, and Mann, *Estimated Use of Water in the United States in 1980*, pp. 22-23.

126. Ibid., p. 26. Fresh surface water accounted for about 71 percent of total cooling-water withdrawals in 1980.

127. Kasper et al., "Use and Cost of Chlorination Systems," pp. 54-57.

128. Robert L. Loftness, *Energy Handbook*, 2d ed. (New York: Van Nostrand Reinhold Co., 1984), table 4-7. p. 121; and Solley, Chase, and Mann, *Estimated Use of Water in the United States in 1980.*

129. Calculated from figures in U.S. Department of Energy, Energy Information Agency, "Thermal Electric Plant Construction Costs and Annual Production Expenditures: 1980," Monthly Energy Review, June 1983.

130. See reference 142 in chapter 3.

131. A 0.1 percent increase in efficiency would provide benefits of $0.71 \times 2.0 \times 10^{12} \times \frac{.001}{.33} \times 2.35$

$$= \$101 \times 10^{9}$$

where:

0.71 = 150/210
= the proportion of thermoelectric cooling water taken from fresh surface-water sources (Waldon R. Kerns et al., "Nonpoint Source Management: A Case Study of Farmers' Opinions and Policy Analysis" [Blacksburg, Va.: Virginia Polytechnic Institute and State University, n.d.], p. 26).

2.0×10^{12} = the total amount of electricity generated by thermoelectric plants in 1980 (U.S. Department of Energy, "Thermal Electric Plant Construction, Costs and Annual Production Expenses: 1980").

.001 = increase in efficiency

.33 = original efficiency

2.35 = cost of producing thermoelectric power in cents per KwH (U.S. Department of Energy, "Thermal Electric Plant Construction, Costs and Annual Production Expenses: 1980").

132. Emerson and Akhavipour, *Benefit Assessment of Kansas Water Quality Management Plan*, p. 74.

133. Reported in Fred H. Abel, Dennis P. Tihansky, and Richard G. Walsh, "National Benefits of Water Pollution Control," prepared for U.S. Environmental Protection Agency, Washington Environmental Research Center, Office of Research and Development, n.d., p. 45.

134. U.S. Department of the Interior, Water and Power Resources Service, Colorado River Water Quality Office, Engineering and Research Center, *Economic Impacts on Agricultural, Municipal, and Industrial Users* (Washington, D.C.: U.S. Government Printing Office, 1980), p. 18.

135. The damages to agriculture for each additional milligram per liter of salinity (in 1976 dollars) is given by the equation $Y = be^{mx}$, where Y = \$ per milligrams per liter, $b = 59.7098537$, e = base of natural logarithms (2.718281828), $m = 0.0051913$, and x = salinity level. (U.S. Department of the Interior, *Economic Impacts on Agricultural, Municipal, and Industrial Users*, p. 9.)

The average salinity level is approximately 900 milligrams per liter (U.S. Department of the Interior, *Colorado River Water Quality Improvement Program*, p. 3.

Thus, the total cost of damage to agriculture is

$$C = \int_0^{900} be^{mx} \, dx$$

$$= \frac{b}{m} \, e^{mx} \Big|_0^{900}$$

$$= \$1{,}220{,}000$$

Converted to 1980 dollars, the total cost is \$1.7 million.

136. Since about 37 percent of the salt comes from irrigation (U.S. Department of the Interior, *Colorado River Water Quality Improvement Program, p. 4)*, the salinity level without irrigation would be $0.63 \times 900 = 567$ milligrams per liter. Therefore, the cost added by the irrigation would be

$$C = \int_{567}^{900} be^{mx} \, dx$$

$$= \frac{b}{m} \, e^{mx} \Big|_{567}^{900}$$

$$\approx \$1 \text{ million}$$

Converted to 1980 dollars, the total cost is \$1.5 million.

137. Based on information from Paul Kossa, U.S. Department of Agriculture, August 5, 1983.

138. Solley, Chase, and Mann, *Estimated Use of Water in the United States in 1980*, p. 18.

139. Nutrient contents based on data in Briggs and Ficke, *Quality of Rivers of the United States, 1975 Water Year*, pp. 35, 39.

140. Clark, "Estimated Effects of Non-point Source Pollution."

6. Techniques for Control

Recognizing that a problem exists is only the first step in the often long process of implementing policies to control it. The obvious next step is asking whether means are available to deal with it. Once that question is answered, others must follow: Are those means effective? Do they control the problem reasonably efficiently? Are they affordable?

The analysis presented in this chapter is far from exhaustive. It does, however, demonstrate that the answers to all these questions are, for most circumstances, yes—the techniques to control soil erosion and runoff do exist, and they are effective, efficient, and affordable.[1]

Water pollution caused by agricultural land runoff can be dealt with in three basic ways. One is to prevent pollutants from being created in the first place. In the case of soil erosion, this means preventing the erosion from occurring and the associated contaminants from being carried off the land. This can be accomplished by tillage and land-management practices such as contour plowing, conservation tillage, and improved crop rotation, by on-field structural measures such as terracing and diversion channels, and by a reduction in the use of pesticides and fertilizers.

The second approach is to attempt to collect and reduce pollutants after they leave a field but before they cause any damage. For agricultural pollutants, this strategy can involve constructing sedimentation basins and filter strips along streams.

The third approach is to allow the problems to occur and to try to compensate for them afterward—for instance, by dredging reservoirs, treating drinking-water supplies, and chemically treating lakes.

The relative efficiency and attractiveness of each of these strategies depends on both the particular situation being addressed and the relative magnitudes of on-farm and off-farm damages. All three approaches probably will be involved in any comprehensive program. Examples of the third approach were discussed in chapter 5 and provided a basis for many of the cost estimates given there. This chapter, therefore, focuses on the first two approaches.

RELATIVE EFFICIENCIES

Three factors must be considered when calculating the relative efficiencies of alternative approaches to erosion control: the direct costs of the techniques, their effectiveness, and any possible ancillary effects of the techniques.

Costs

Techniques for controlling the discharge of sediment and associated contaminants into waterways can create several types of costs. The most obvious is the cost of the resources—the labor, energy, machinery, and other inputs—consumed in implementing the technique. Some techniques require the purchase of special machinery; some require additional labor, either in the investment stage (for instance, when a structure is built) or in the operating stage (as in the additional labor required for some soil-conserving cultivation techniques); and some may require increased inputs such as fertilizer and pesticides to maintain existing crop yields.

A second type of cost is the loss in crop-production value that can result from the adoption of erosion-control techniques. Some techniques may lower the yields of existing crops; some may involve planting lower-value crops (for instance, alfalfa instead of sugar beets, or clover instead of corn); and some may take formerly cropped land out of production entirely.

Another type of cost, often more difficult to quantify, is increased risk and uncertainty. In some cases, the new techniques may increase the risk of low yields. Many farmers manage their farms very conservatively—tilling the fields as early as they can, applying excess amounts of fertilizer and pesticides—to keep such risks at a minimum. Management schemes that involve less conservative farm practices may well increase such risks.

Increased uncertainty will usually accompany new techniques even when they do not increase the risk of low yields. Often neither the farmer nor society has had sufficient experience with a new technique to be sure how it will perform on a particular field under the specific conditions that are likely to occur. Such knowledge, and its associated certainty, come only with experience.

A final cost consideration is the possible need for improved managerial skills. Some of the techniques may require more knowledge and care on the part of the farmer (such as learning to apply pesticides only when they are truly needed rather than according to a fixed schedule). The costs implied in the attainment of these skills are very hard to quantify but may nevertheless be important.

Increased inputs, lost revenues, higher risk and uncertainty, and greater skill all pertain to most techniques. However, they can function as benefits as well as costs. Some techniques result in a reduction in the inputs required, an increase in crop production, or reduced risk. Further, there often is a trade-off between one cost category and another. Increased managerial skill and labor input in improved pesticide application, for instance, can result in lower chemical costs and higher yields as well as lower pesticide runoff from fields.[2]

Effectiveness

The effectiveness of a technique should be measured, of course, by how well it reduces any off-site problems of concern. Such a measure, however, necessarily depends on the specific location in which the technique is being proposed and the specific problems that are being addressed. Physical, chemical, and biological variables such as soil type, crops grown, field slope, rainfall patterns, climate, and distance from streams will determine how effective a particular technique is on a particular farm.

Some techniques only control some problems. Sedimentation basins, for instance, can be quite effective in controlling the amount of sediment and other solid particles reaching a waterway but may do little to reduce the amount of dissolved contaminants. Thus, the effectiveness of a technique must be measured in terms of the specific problems that are to be controlled.

These considerations make any general assessment of effectiveness very difficult, at best. This chapter only summarizes the results of selected experiments and studies that provide some insight into how effective the different techniques may be. It is intended only to be indicative, not comprehensive.

Ancillary Effects

When assessing the relative desirability of different types of techniques, it is important to consider their possible ancillary effects. These can be either positive or negative. For instance, some techniques may provide such benefits as additional wildlife habitat, water storage (for use during dry periods and for watering livestock), or groundwater recharge.[3] In some cases, however, a technique may only move a problem somewhere else—such as when contaminants are allowed to seep into groundwater supplies rather than run off fields into surface water.

As with effectiveness, the importance of ancillary effects depends on the specific conditions under which a new technique is adopted.

TECHNIQUES

Substantial discussion and analysis of various techniques for controlling soil erosion, along with handbooks for their adoption, are available elsewhere.[4] This section briefly summarizes the relevant aspects of the different techniques, and, to the extent available information allows, it analyzes the costs, effectiveness, and ancillary effects of various types of techniques. The first two types discussed—tillage practices and cropping patterns—primarily operate by preventing pollutants from leaving a field. By contrast, although some structural measures and other land-management practices operate similarly, others act by intercepting pollutants before they cause problems.

Tillage Practices

Chapter 2 noted that tillage, or the way in which a farmer prepares a field for planting and controlling weeds, can significantly affect the amount of erosion that occurs and the amount of soil and associated contaminants carried off the field. Improved tillage practices have long been the mainstay of soil conservation programs, and their preeminence can be expected to continue.

Contouring

Contouring, one of the longest used conservation techniques in the United States, involves plowing, planting, and harvesting along the contours of hills, rather than straight up and down their slopes. The furrows catch and hold water, allowing it to seep into the ground and thereby reducing runoff.

Contouring often costs only a few dollars an acre. The major expenses include additional labor, time, and managerial skill required to plow according to the field topography. These costs, however, can increase significantly where there is highly variable topography and when the farmer is using large, wide-swath machinery.[5] Recent increases in machinery size, farm size, and labor costs have reduced the frequency of contour plowing in the United States.

Under the right conditions, contouring can substantially reduce sediment, nutrient, and pesticide losses.[6] A study sponsored by the U.S. Environmental Protection Agency (EPA) estimated that contouring reduces suspended sediment (and other suspended solids) by about 20 percent,[7] but other studies have found reductions of up to 75 percent, with many estimates falling in the range of 25 to 50 percent.[8] It is likely to be most effective under two of the conditions—moderate slopes and topography—that make it inexpensive.[9] Contouring's effectiveness also is increased by moderate rainfall intensity, high rates of water infiltration, large soil-particle sizes, and more cohesive soils.[10]

Contouring can reduce dissolved nitrogen runoff by 25 to 50 percent and both nitrogen and phosphorus in solid phase by 55 to 65 percent.[11] Depending upon the particular pesticide being used, pesticide runoff can be reduced 20 to 25 percent[12] and total annual runoff by up to 20 percent.[13]

Conservation Tillage

Conservation tillage is an array of reduced-cultivation practices that protect soil by leaving a mulch layer of residue on the surface. Its fast-growing popularity (24 percent of all U.S. cropland in 1982[14]) is attributable both to its demonstrated effectiveness in controlling erosion and runoff and to its ability to reduce production costs.

Conservation-tillage practices cover a broad spectrum varying in the amount and type of tillage done. The most "extreme" form is "no-till," in which seeds are planted directly through a previous crop's residue. In this practice, at least 90 percent of a crop's residue remains undisturbed on the soil surface. "Mulch-till" and "reduced-till" involve some conventional tillage. They typically leave 20 to 30 percent of the crop residue on the surface and incorporate the rest into the soil. With "ridge-till," seeds are planted on ridges between furrows filled with crop residue. Tillage is limited to a narrow strip on which seeds are planted with "strip-till."[15] Most of these tillage systems depend on herbicides to control weed growth, but there is disagreement on whether these systems necessarily require more herbicide use than conventional tillage does and, if they do, whether they result in increased herbicide runoff.[16]

In general, conservation-tillage practices are inexpensive. EPA estimates a cost of $31 per acre,[17] although the International Joint Commission estimates the cost to be lower—only $1.50 to $7.60 per acre.[18] In many cases, they result in reduced cultivation costs because of labor and energy savings, although they usually require a farmer to purchase special equipment.[19] In addition, expenditures on herbicides and insecticides can increase, particularly if a weed species becomes resistant to the herbicides used or if the mulch provides habitat for other pests. Several studies estimate cost savings of up to $25 per acre for no-till,[20] but one study focusing on corn production found that increased pesticide expenditures required under conservation tillage largely offset the decreased expenses for labor, energy, and machinery use.[21] Nevertheless, some persons have argued that weed problems are no greater with conservation tillage than with conventional systems (although the techniques of responding to the problems can differ).[22] A need for increased pesticide expenditures depends partly on what crop is planted—the costs are typically higher for corn than for other crops.[23]

The evidence of conservation tillage's effect on crop yields is mixed. Yields may be lower in northern areas and in areas with finely textured, poorly drained soils where surface mulch slows soil warming and drying, delaying spring planting.[24] The mulch also may immobilize surface-applied fertilizers and harbor pests that reduce crop yields.[25] In dry climates with well-drained soils, however, several studies have found that no-till consistently produces higher yields.[26] Two possible reasons are that the mulch reduces soil moisture evaporation and that it promotes the growth of bacteria that break down organic nitrogen compounds, making the nutrient more available for plant growth.[27] Not surprisingly, whether conservation-tillage practices result in an increase or decrease in net farm profit is determined by the impact they have on crop yields.[28]

Conservation-tillage practices can be very effective at limiting the amounts of runoff, sediment, and nutrients that are lost from a field. This effectiveness seems generally proportional to the amount of residue left on the field.[29] Runoff can decrease by an average of 25 percent according to one study,[30] but other studies show reductions of 10 to 60 percent, depending on the soil type, method of conservation tillage, and climate.[31] Sediment losses can be reduced by 15 to 90 percent,[32] and nutrient losses by 15 to 70 percent.[33] However, estimating nutrient losses can be tricky—for instance, vegetative cover can sometimes increase the levels of soluble nutrients in runoff (see chapter 2) while decreasing losses of insoluble nutrients. Moreover, there is no clear evidence of what effect conservation tillage has on pesticide runoff, an important consideration since such tillage often results in increased pesticide use.[34] EPA estimates that herbicide losses can be reduced up to 99 percent,[35] but other studies often show the same or higher losses with conservation tillage.[36] Even if pesticide runoff is reduced, its concentration in the runoff water may be twice as high.[37] And, if conservation tillage does reduce pesticide runoff, it may have a related ancillary cost: because these tillage practices are typically associated with higher rates of water infiltration, increased herbicide use may also mean increased groundwater contamination, especially in sandy soils.[38]

Because conservation tillage changes the physical, organic, microbiological, and chemical environments in soils, farming practices such as fertilizer management should also be adjusted.[39] Similarly, a need to reduce herbicide runoff will complicate decisions on what herbicide is best for a specific situation and when, how, and at what level it should be applied.

Other Tillage Practices

Various other changes in tillage practices can also reduce the amount of sediment and other contaminants carried off farmland. In general, substituting chemical for mechanical weed control leaves residues on the surface that can protect soil from rainfall and can retard runoff. If a soil surface tends to form a hardpan, however, mechanical weed control may increase the amount of water that seeps into the soil, also reducing runoff.

The timing of any tillage undertaken can be important. For instance, delaying plowing until the spring rather than plowing crop residue under soon after harvest can significantly reduce the amount of erosion that occurs during the winter and early spring. In some areas, however, such a delay may slow the rate at which the soil dries out and warms up in the spring, delaying spring planting and thereby reducing yields.[40]

The costs of such changes in tillage practices are usually small, and they can, in certain circumstances, result in large reductions in erosion and runoff.[41] Although these tillage practices are likely to be less effective than contouring or conservation tillage at controlling erosion, they can serve as useful supplements to those techniques.

Cropping Patterns

As indicated in chapter 2, one of the most important variables affecting the rate of erosion is the type of crop grown on the land. Row crops typically leave large amounts of land uncovered and allow water to flow easily down the rows and off the field. Field crops, however, provide more of a canopy and retard runoff. In some cases, removing a field from row crops may be the only way to reduce erosion effectively and efficiently. In many cases, however, substantial improvements can be achieved with less dramatic shifts.

Crop Rotation

Farmers have used crop rotation—that is, they have included soil conserving crops in the sequence of crops grown on a field—as a soil conservation measure for centuries. Some crops (for instance, corn) are soil depleting: they provide little protection against erosion and have high nutrient demands. Others, such as some grasses and legumes, return nitrogen and organic matter to the soil, improving its nutrient levels, and provide a thick, protective canopy against erosion while they are growing.

Rotating such crops can result in substantial savings in production costs. The soil-building crops decrease the amount of fertilizer that has to be applied to the soil-depleting crops later in the rotation.[42] Such rotations can also disrupt weed, disease, and insect cycles, reducing the need for high pesticide applications. For instance, corn rootworm often can be completely controlled by using an appropriate crop rotation.[43]

One crop-rotation study—using a six-year rotation pattern with three years corn, one year oats, and two years hay—found a savings of $47 per hectare in rootworm insecticide and nitrogen costs for the first year following hay.[44] Rotations can, however, decrease farm income because soil-building crops often produce lower yields or sell at lower prices than do soil-depleting crops.[45] One study estimated an average cost of $36 per acre per year in lost income compared to continuous corn cropping.[46]

Crop rotations have been found to reduce phosphorus runoff by 30 to 75 percent and nitrogen runoff by 55 to 80 percent.[47] Although a crop rotation that includes meadow or grass has the greatest potential for reducing erosion, rotating between grain and row crops can also result in reductions of 50 to 70 percent.[48] In one Missouri case study, average annual erosion rates of 19.7 tons of soil per acre with continuous corn cultivation dropped to 2.7 tons per acre when a corn, wheat, and clover rotation was used.[49] Rotation's impact on pesticide loss can be even greater because it can reduce the amount of a pesticide used as well as the proportion that leaves the field.[50]

The ancillary effects of crop rotation generally appear to be positive. It can increase groundwater recharge with less concern about groundwater contamination (because less fertilizer and pesticide need to be applied to the land).* It also can provide increased habitat for wildlife

---

*Groundwater recharge may increase for two reasons. For one, by restricting the rate at which water runs off a field, field crops provide more opportunity for water to seep into the ground during the periods in the rotation when such crops are being grown. The second is that the soil-conserving crops usually improve soil texture, allowing rainfall to be absorbed faster even while row crops are growing in subsequent seasons during the rotation.

and can decrease the risk, which exists in a continuous monoculture cropping system, of disease or insect epidemics.

Cover-Cropping

Cover crops, usually close-growing grasses or legumes planted when the land would otherwise be left fallow, can both temporarily protect the land from erosion and provide the plant residues necessary for conservation tillage when the land is once again brought into production. Cover-cropping already is a popular practice: 40 percent of Agricultural Conservation Program funds in 1981 and 1982 went to fund cover-crop establishment and improvement.[51]

Cover crops involve additional costs for seed, labor, and machinery operation, but these are generally small. In addition, the crops may lengthen a field's drying time in the spring and delay spring tillage, lowering yield in the next crop.[52] However, by fixing nitrogen and contributing organic material to the soil, such crops may also reduce the amount of fertilizer that must be purchased and increase yields for crops grown later. Moreover, farmers may be able to market some of the cover crops or use them as forage.

Cover crops effectively reduce rainfall impact and runoff while they are growing.[53] They may also reduce the leaching of nitrates from the soil.[54] Two ancillary benefits are the food and habitat they provide for wildlife and their tendency to retard the runoff of melting snow in the spring.

Strip-Cropping

Strip-cropping entails planting strips of close-growing crops such as alfalfa and meadow grasses as buffers between strips of row crops such as corn. The direct costs are moderate, limited to some seed and planting expenses. An EPA-sponsored study concludes that the direct crop-production costs can fall $30 per acre with the adoption of strip-cropping because of savings in fertilizer and pesticide applications and in equipment use, with only labor costs rising slightly.[55] However, the value of the crops raised is also likely to be lower on a per-acre basis—about $70 an acre in the EPA-sponsored study.[56]* The strips

---

*Comparisons that focus on the cost per acre, however, may not be appropriate in determining the relative cost of strip-cropping. In many cases, the relevant comparison is whether total production costs and value of output are significantly affected for an entire farm unit raising both row crops and cover crops if they are planted in alternating strips rather than on separate fields.

may also require increased skill and labor for tillage and harvesting and become more troublesome when larger farm machinery is used for cultivation and harvesting. The International Joint Commission estimated that the total costs of strip-cropping amount to about $16 per acre.[57]

Strip-cropping's effectiveness primarily results from its breaking the length of a slope into segments, reducing runoff velocities, and filtering sediment and trapping nutrients carried off the row crops. It is most effective when the strips are planted along contours and the hill slopes are not too steep. This widely used technique can reduce soil loss by up to 85 percent.[58] There is little information about the effectiveness of strip-cropping by itself in reducing runoff and fertilizer and pesticide losses because it is often adopted in association with other techniques such as contouring and crop rotations. Strip-cropping on the contour, however, can reduce sediment and yields by 75 percent, perhaps 25 percent more than contour plowing alone.[59]

Intercropping

Intercropping is a form of cover-cropping but refers to planting low-growing crops as undergrowth for orchard or other crops. Again, the low-growing crops used commonly are legumes or grassy species that can add nutrients to the soil and can reduce erosion by providing a protective second-layer canopy and lessening surface flow.

Structural Measures

A third major category of erosion-control measures involves creating structures that retain or redirect runoff water. Such structures include terraces, diversion channels, sediment basins, and grassed waterways that can be effective for many years if they are properly maintained. Terraces and diversion channels reduce on-field erosion, while sediment basins and grassed waterways reduce sediment delivered to receiving waters.

Terraces

Terraces are earthen embankments constructed across the slope on hillsides to divide a slope into a series of steps. By reducing the length of a slope and intercepting water flowing off a hillside, they reduce the water's erosion potential and allow any soil that it has picked up to be redeposited.

Terraces are one of the most effective ways to reduce sediment delivery nutrient loss and pesticide runoff. In reducing the amount of water runoff (by up to almost 90 percent), the amount of soil loss (in most cases, by 75 to 95 percent), and the amount of resulting suspended sediment (by 30 to 50 percent according to EPA and by 75 percent according to another source), they catch both dissolved and adsorbed particles.[60] As a result, total nitrogen losses can be reduced 80 to 90 percent according to one study and 25 to 95 percent according to another.[61] According to the same studies, phosphorus loadings are reduced 56 to 89 percent and 20 to 95 percent. Pesticide loadings may be cut in half.[62] Terraces are most effective, however, on gradually sloping terrain. On land with a slope greater than 12 percent, erosion from the steep back slopes that form the terrace tends to negate the benefits of the terraces.[63]

Like most soil conservation measures, a major ancillary benefit of terraces is that they conserve water, since it seeps into the ground rather than runs off the surface.[64] This is probably why many terraces are constructed in the Great Plains or on other land with only a moderate slope (two-thirds of the terraced cropland in the United States is located in fields with a natural slope of less than 4 percent) and where the erosion potential is very low.[65] In addition, most terraces are located in areas where rainfall is sparse and water conservation is particularly important.[66]

Terraces do have their drawbacks. For one thing, they can complicate farm management. In fact, many of the terraces built in the past are being torn out because they are incompatible with the large, heavy machinery being used increasingly in U.S. agriculture.[67]

Terraces also are expensive, particularly in their initial construction cost. The initial construction cost can amount to $500 to $900 per acre, and the amortized construction costs can amount to $25 to $41 per acre according to one study and $50 to $100 per acre according to another.[68] EPA estimates that the installation costs average $73 per acre, and annual maintainence costs $16.[69] However, their cost per acre will vary, depending on how many terraces are constructed on a given slope.[70] A soil conservation study in Iowa argued that terraces should only be used as a last resort for reducing soil loss because of their expense and the inconvenience in farming around them.[71]

Diversion Channels

Diversions are vegetated channels constructed across the slope of a field to catch water and carry it off the field. They usually are not costly to construct—one source estimates a annualized cost of $6 per acre—but must be regularly maintained by mowing the vegetation and periodically removing accumulated sediments.[72] Their primary purpose is to divert water off the field, thereby preventing downslope erosion. Since they require that some land be diverted from crop production, they may impose additional costs on the landowner, estimated to be $14 per acre in an EPA-sponsored study.[73]

Little information is available about the success of diversion channels in reducing sediment and other contaminants in runoff water. In some cases, the channels themselves can start to erode, creating a potential for gullying.

Sediment Basins

Sediment basins are catchments designed to impound runoff water long enough for suspended sediments to settle out. They can be constructed along a stream or between a field and a waterway.[74]

Such basins can be expensive to build—upwards of $2,000 each—and they require continual efforts to remove and dispose of accumulated sediments and maintain the structure.[75] Since they are constructed off a field, they have no effect on crop yields.

Sediment basins can be very effective in capturing sediment—one study estimates that they capture up to 90 percent of total suspended solids[76]—and adsorbed particles but are less effective in controlling dissolved contaminants. They can reduce total nitrogen and phosphorus loadings by about 50 percent.[77] Their effectiveness depends on the volume of water they can store compared to the volume that flows into them during a storm. The larger the ratio of storage volume to inflow, the more effective the basins will be. If the ratio is small, they may capture only the larger particles, allowing fine particles, with their adsorbed nutrients and other contaminants, to flow through.[78]

These basins can have several ancillary effects. First, they can reduce flood peaks, thereby reducing erosion downstream and increasing the stability of a stream channel.* If the basins cause water to remain in ponds for long periods of time, they can also provide an additional water source for livestock or for withdrawal during droughts and can support riparian wildlife habitat.[79]

Grassed Waterways

Grassed waterways are natural or constructed channels lined with vegetation to prevent gully formation. They are effective even on steep

*However, because of its reduced sediment load, water leaving the basin will be more erosive than water entering it.

land or when lined with grasses with relative low tolerance of frequent inundation.[80] The cost of building and maintaining such waterways is often less than other structural measures, although they may take some acreage out of crop production and interfere with the use of large machinery.[81]

In addition to preventing gully formation, the vegetation decreases runoff volume (by 2 to 50 percent according to one study[82]) by facilitating infiltration[83] and traps sediment particles (an estimated 25 percent according to one source[84]) in the runoff, although other measures (for instance, sediment basins) installed specifically for trapping sediment are usually more efficient at that purpose.[85] Grassed waterways can also capture some of the nutrients and pesticides in the runoff water. One study found nitrogen runoff reduced by about 70 percent,[86] while another estimated that nutrient losses would be reduced only 10 percent.[87] In a pesticide-transport study, only 30 percent of the pesticide 2,4-D carried off a field was transported to the end of a 25-yard-long grassed waterway. In that study, the vegetation itself, rather than simple infiltration, seemed responsible for the control of 2,4-D delivery since comparable reductions in pesticide concentrations were observed regardless of how saturated the soil was prior to the runoff event.[88]

Other Land-Management Practices

A variety of other land-management practices can contribute to controlling soil erosion and reducing the runoff of nutrients and pesticides. Attempting a complete list of such techniques is beyond the scope of this chapter, but some of the more important ones are briefly described below.

Filter Strips

Filter strips (also known as buffer strips) are strips of close-growing vegetation planted along the downslope edges of cultivated fields or between the fields and adjacent streams. Such strips can cost relatively little to establish—the International Joint Commission estimates costs of $100 to $1,200 per acre of land included in the strip[89]—but may take valuable land out of production. They also may need to be maintained if they are located in areas highly susceptible to erosion. Filter strips reduce runoff velocity, thereby allowing sediment deposition and runoff infiltration; different sources estimate that they reduce both suspended sediment loadings and runoff by 25 percent.[90] They also filter out pesticides, nutrients (one study estimates a 52 percent reduction for nitrogen and 24 percent reduction for phosphorus, while another estimates about 50 percent reduction for both[91]), and microorganisms. Their effectiveness is influenced by their width relative to runoff volume, density of grass cover, detention time of water flow in the strips, and uniformity of flow over the strips.[92]

An ancillary benefit is that filter strips can provide valuable riparian wildlife habitat and may help stabilize a stream channel if they extend all the way to the bank.

Mulching

Mulching is a general term referring to the use of plant residues (such as straw or shredded corn stalks) or other materials (such as compost or wood chips) as a protective cover on erodible hillsides and cropland. It is used primarily to control the spots in a field where soil loss is particularly likely (for instance, a steep slope on a field's edge or a gully through a field). Mulching can be used as temporary protection until a grassy crop (which is less expensive to maintain) takes hold.[93] Mulching not only reduces erosion (decreasing total suspended solids by 50 percent according to one study[94]) and decreases the runoff of associated contaminants (by 25 percent according to the same study) but sometimes also increases soil moisture and decreases weed growth. The cost of mulching depends on the cost of the material used and the difficulty of spreading it. A Tennessee planning report estimated an average cost of $310 per acre.[95]

Retiring Highly Erodible Cropland

Sometimes, there may be no economical way to prevent serious erosion while keeping land in row crops. This may be the case particularly with erodible land on steep slopes. In such cases, the most reasonable action may be to convert the land to growing field crops such as alfalfa or grains, to use it as pasture for livestock, or to take it out of production altogether. The U.S. Department of Agriculture encourages these actions in special cases with a conservation reserve program that pays farmers to take land out of crop production. In 1984, this program was expected to result in over a quarter of a million acres of highly erodible land being taken out of crop production.[96]

The primary cost of such actions is the value of the lost production. However, production costs on such lands can be high and yields low, in part because fertilizers easily wash off the land in a storm. As a result, if the land is planted in forage or grain crops, the loss of income to the farm may not be significant.

If the alternative vegetation becomes well established on the land, erosion and associated problems can be reduced substantially.* However, even if a field is taken out of production, it usually is necessary to plant it with grasses to ensure that this cover develops properly and that the ground is not left unprotected during the sometimes slow process of natural revegetation.

As an ancillary benefit, allowing land to revert to its natural vegetation can result in the creation of valuable wildlife habitat.

Pesticide- and Fertilizer-Management Practices

Erosion-control practices are not always sufficient for fertilizer and pesticide control since nutrients and pesticides often are lost through surface runoff, seepage through the ground, or conveyance off a field by breezes while they are being applied. Perhaps the most effective methods of controlling pesticide and fertilizer pollution are preventive measures: reducing either the amounts of pesticide or fertilizer initially applied or their vulnerability to runoff. Careful management of chemical applications is especially important in conjunction with erosion-control measures that increase infiltration (particularly in sandy soils) to protect against groundwater contamination.

Incorporating fertilizers and pesticides into the soil, rather than simply spreading them over the surface, is one way to significantly reduce runoff losses. Researchers have estimated that the layer of soil and chemicals that mixes with rainfall and runoff water is extremely thin—perhaps only six millimeters thick.[98] Also, the rate at which phosphates convert to insoluble forms can be increased by fertilizer incorporation.[99]

Of the several possible incorporation methods, placing subsurface bands of fertilizer along seed rows is one of the most efficient.[100] It places fairly immobile phosphorus fertilizers in direct proximity to plant root systems, where the nutrients will be most readily available, and minimizes losses due to surface erosion.[101]

Pesticides, too, can be incorporated below the soil surface whenever appropriate. It has been shown that surface runoff losses of three different herbicides disked into the soil were only one-third what they would have been using surface application without incorporation.[102] However, improved soil retention of fertilizers and pesticides also can have a major ancillary cost—increased groundwater contamination.

*Nevertheless, one author has cautioned that "uncultivated lands can yield appreciable losses of nutrients."[97]

The timing of fertilizer and pesticide application can be particularly important for easily leached materials such as nitrogen fertilizers. Improving the timing of fertilizer application almost always saves money. By applying the chemicals when they will be most effective, overall chemical use is reduced and crop yields improve.[103] One study estimates that improved fertilizer and pesticide management can reduce runoff of both by up to 50 percent,[104] while the International Joint Commission estimates that pesticide runoff can be reduced by up to 75 percent through better management.[105]

The application of pesticides, and thus their losses in runoff, can also be reduced by using integrated pest management (IPM), a range of ecologically based management strategies that combine biological controls (use of pests' natural enemies and pest sterilization), crop rotations that discourage infestation, planting of resistant crop strains, and selective, carefully targeted applications of chemical pesticides.[106] Under IPM, instead of increasing the application of pesticides according to a fixed schedule without regard for whether they are needed, a farmer assesses the economic threshold at which the benefits of pest control outweigh the value of crops lost to pests.[107]

To be effective, these pest-management strategies require improved management skills and more labor. They may also, in the short run, increase risk.[108] However, they reduce pesticide costs substantially and can result in long-term increases in yields. One study has predicted that yields can be expected to increase 10 to 25 percent in the future as a result of IPM strategies.[109]

EPA estimates that IPM alone will reduce pesticide runoff 20 to 40 percent,[110] while an EPA-sponsored study estimates that, although total pesticide use could be reduced by 40 percent with the most economical IPM currently available, that total reduction could increase to almost 60 percent with methods likely to be available within 5 to 10 years.[111] IPM has potentially significant ancillary benefits in reducing human pesticide exposure (to mixers, applicators, and consumers of crops grown) and decreasing pesticide risks to birds and other wildlife.

Controls on Irrigated Fields

Controlling erosion and runoff from irrigated fields may involve management practices unique to irrigation farming, as well as those used in rain-fed areas. Even in areas where there is little erosion, irrigation return flows can be laden with dissolved fertilizers, pesticides, and salts.[112]

The major cause of erosion on irrigated cropland is furrow irrigation of sloping lands,[113] but, by exercizing greater care in making the furrows, that erosion can be reduced. An Idaho study found that reducing furrow size by about half not only reduced erosion but apparently resulted in the capture of sediment and phosphorus carried onto the fields by irrigation water.[114]

Other techniques for reducing furrow-caused erosion involve reducing the capacity of the furrows as they reach the end of a field (leading to declines in sediment loss of up to 93 percent), irrigating in non-compacted furrows, recycling tailwater either to the same field or to other fields with close-growing cover crops such as alfalfa (in one such instance, 79 percent of the sediment in runoff water was removed), and restricting preplant irrigation to seed beds rather than saturating an entire field.[115] All these measures are relatively inexpensive; some may even reduce production costs.

In addition, the use of alternative irrigation techniques such as sprinkler or drip systems can reduce erosion and runoff, although they can be quite expensive to install and operate.[116] Such systems also require greater managerial skill and labor input than traditional flood or furrow irrigation. However, in addition to reducing off-farm impacts, sprinklers or drip irrigation substantially reduce water consumption.

Many of the other already-described off-field control techniques are also effective for irrigated land. Filter strips planted along the lower edge of an irrigated cornfield have been found to filter approximately 45 percent of the sediment from irrigated runoff.[117] Small sediment basins that catch and detain runoff from several adjacent furrows can remove up to 95 percent of the sediment, and sediment ponds either on a field or along waterways have sediment-removal efficiencies of between 40 and 70 percent.[118]

MAKING CHOICES

No one of these erosion-control techniques, or their many variations, will be suitable for all situations. With the thousands of different soil types, combined with the variety of other factors such as rainfall patterns, cropping patterns, and topography, there is an infinite variety of conditions that determines the efficacy of any particular erosion-control technique. Moreover, despite the substantial research and analysis that have been done on these techniques, much still needs to be learned about individual conservation systems.[119]

Choosing Among Techniques

Nevertheless, enough information has been collected about the costs and the effectiveness of the different techniques to allow choices to be made among them. Tillage practices such as conservation tillage and contour plowing are less expensive than structural measures such as terraces and sediment basins. In addition, the high construction costs of terraces and sediment basins are not offset by savings in inputs such as fertilizers. However, with conservation tillage, stripcropping, and other techniques based on tillage practices and cropping patterns, the critical economic factor is yield changes. One study concluded that decreased expenditures for labor and machinery with conservation tillage almost balance increased pesticide expenditures; thus, the variable that determines changes in farm income is the effect of conservation tillage on crop yields.[120]*

Terraces, conservation tillage, and contour plowing are the most effective conservation techniques at controlling on-field sediment loss, while sediment basins are the most efficient in controlling sediment loss to waterways. For all techniques, the higher the sediment-delivery ratio, the larger the percentage reduction in sediment loadings is likely to be after the erosion-control system is adopted.[121] Nevertheless, the effectiveness of any technique depends on a variety of site-specific characteristics such as soil type, field slope, climate, and rainfall patterns. A study of "best management practices" (BMPs) in the Palouse River basin concluded that mulching yielded the greatest return per acre in low-precipitation zones, while conservation-tillage returns were greatest in high-precipitation zones.[122]

Soil-detachment-control practices such as conservation tillage are apparently less effective at preventing movement of both dissolved and adsorbed nutrients than are transport-control practices such as terraces and contour plowing. The decreased sediment and runoff losses associated with terraces and contour plowing also reduce nutrient losses. Although all the erosion-control practices discussed in this chapter reduce both dissolved and solid nutrient losses, crop rotation is the only practice other than improved fertilizer management that also decreases the amount of dissolved nitrogen that percolates into the groundwater.[123] Although other practices can also enhance soil moisture content, they may simultaneously increase nitrogen concentrations in groundwater.

The ability of a management practice to control pesticide losses depends largely on the particular pesticide being applied and the

*The level of pesticide expenditures is also highly variable, with less required where soils are coarse and well drained.

method of application. By targeting sediment and runoff losses, rather than pesticides, many management practices do not control a large fraction of pesticide losses. Although the data on the fate of pesticides after field application are sketchy, it would appear that integrated pest management, because it substantially reduces the amount of a pesticide applied, is the most effective control practice.

A study by the Soil Conservation Service of the U.S. Department of Agriculture surveyed the costs and effectiveness (in terms of reduced erosion) of different management practices and different pretreatment erosion rates. Strip-cropping, terraces, and conservation tillage were found to control the largest amount of sediment per dollar spent. The least efficient measure was interim cover (cover crops planted during the noncrop season).[124]* With all the conservation techniques included in the survey, an increase in an initial erosion level was accompanied by a substantial decrease in the cost of removing a ton of sediment—but the relative rankings of the different techniques changed little. Other studies have also shown that, as sediment-delivery ratios decrease, the cost per pound of controlled pollutant increases.[126]

All these analyses have been based on the private costs of the erosion-control techniques to individual farmers. However, in planning which erosion-control techniques are most desirable for the entire United States, it is also necessary to consider whether these costs accurately reflect the cost of the techniques to society as a whole. Several factors can cause social costs to differ from private costs. One is the taxes that may be included in some of the private costs but that are not a cost to society because they do not represent the consumption of real resources.†

A more significant factor that also may cause social costs to be lower than a farmer's costs is lost crop production. Although to the individual farmer, this is one of the larger economic effects of some erosion-control techniques, it is not necessarily a cost to a society frequently experiencing excess production. Some of the lost production may actually represent a savings to society because it reduces the amount of grain that the federal government has to buy and store to maintain crop prices. None of the analyses of soil conservation techniques re-

---

*Findings of studies will vary, depending on methodology and site characteristics. Other studies have found terracing to be one of the least efficient sediment-control practices (because of the costs involved in construction), and diversion channels to be one of the most efficient.[125]

†However, taxes are likely to be only a small part of a farmer's expenses.

viewed for this study explicitly considered this factor. To do so might well change the conclusions about which techniques are the most economically efficient.

Moreover, the total economic costs of a whole program to reduce off-site impacts could differ from the sum of individual techniques applied on individual fields. For instance, substantial savings could result if, instead of requiring every farm to adopt control techniques, an entire watershed were treated as a control unit and the most efficient measures were adopted for the watershed as a whole.[127] One study concluded that "appreciable reductions in gross soil erosion can be made with only modest reductions in gross watershed income" by simply shifting cropping patterns within a watershed, assigning crops to different fields on the basis of soil and land characteristics.[128]

In addition, several macroeconomic analyses have attempted to predict the impact of aggressive national erosion-control programs on agricultural prices and income.[129] Some of these studies have indicated that some programs may force consumers to pay most of the costs, with the agricultural sector actually experiencing a net increase in profits. This can be the case, for instance, if a program sufficiently reduces crop production to cause an increase in agricultural prices. The increased revenues resulting from the higher prices may be greater than the combined cost of the control measures and the value of lost production. However, even in these cases, an individual farmer required to adopt the control measures may still be financially worse off.

Choosing Sets of Techniques

Looking at the relative costs and effectiveness of individual erosion-control techniques may not provide insight into what combinations of techniques would be most effective at reducing sediment and other loadings to desired levels.

Although several planning agencies have identified BMPs that are effective for specific situations, there has been very little rigorous analysis of the costs and effectiveness of alternative sets of BMPs. The law of diminishing returns operates with all erosion-control techniques, both when additional techniques are implemented on one field and when additional lands are brought under control on a farm or within a watershed. One study found that marginal control costs increased sharply after the amount of sediment was reduced by more than 70 percent.[130] Another nationwide analysis attempted to measure the importance of diminishing returns in the effectiveness of erosion-control techniques.[131] It found that, when only conventional control measures (for instance, terraces, conservation tillage, and contour plowing) are used, soil erosion routes still increase rapidly when highly erosive land is brought into production. If the control strategy also includes the retiring of particularly erodible lands from row-crop production (see chapter 7), there is still an eventual sharp increase in the average per-acre cost, but many more acres are controlled before that rise begins. As a result, including a public-purchase option results in a much more cost-effective strategy.

Some Broader Choices

The land-management techniques summarized in this chapter have focused primarily on on-farm soil-conservation measures. However, a broader strategy for reducing off-farm impacts could also include two other approaches: adopting measures to prevent sediment and associated contaminants from causing impacts even if they do leave a farm and adopting measures that compensate for the impacts once they have occurred. It is unlikely that a strategy involving either one of these approaches on its own would be very efficient, just as using only on-farm techniques probably could not result in maximal efficiency. However, no work has been done to attempt to identify the most efficient mix of different approaches.

It is also not clear whether controlling agricultural nonpoint sources of pollution is the most efficient way to reduce the environmental impacts described in this book. Most of these impacts are caused by pollutants from several sources, not just agricultural lands. For instance, heavy nutrient loadings to water are also provided by municipal sewage, and some of those nutrients are more biologically available than are those in runoff from farmland.[132]

Could controlling these other sources alone adequately reduce the problems described in this book, and could those sources be controlled more economically than agricultural sources? There is increasing evidence that the answer to both questions is no. As point sources of pollution have been cleaned up in the United States, nonpoint sources have accounted for an increasingly large percentage of pollutant loadings. For those waters that still experience problems after the current phase of point-source abatement is completed, controlling nonpoint sources may be the only alternative available.

Even if additional point-source controls were able to do the job on their own, controlling nonpoint sources could be significantly less costly. An extensive study of pollution problems in the Chesapeake Bay showed that the least-cost strategy for most effectively reducing nutrient loadings to the bay would be to control agricultural runoff by using conservation tillage.[133] Additional treatment of municipal wastewater was substantially less cost-effective. For instance, additional treatment of municipal effluent discharged to the Rappahannock River basin would remove phosphorus from the river (which flows into the bay) at a cost of approximately six dollars per pound, while conservation tillage would remove the nutrient at a cost of less than $1 per pound. Additional treatment of municipal effluent would remove nitrogen for $3.28 a pound, while nonpoint-source controls would remove a pound of nitrogen for about 21 cents.[134]

Further Considerations

Economic efficiency is one of several considerations that need to be taken into account in developing a program to control the off-farm impacts of soil erosion. Equity, political acceptability, and legality are others. These considerations are discussed further in chapter 7.

Recognition must also be given to the ancillary effects of different techniques. Most techniques increase soil moisture and groundwater recharge. Some provide beneficial wildlife habitat and for that reason may be preferred over those that do not.[135] And some will have a more beneficial impact on on-farm productivity than others. These are only some of the factors that have to be considered in identifying what techniques should be implemented. Some of these also are addressed in chapter 7.

REFERENCES

1. See, generally, North Carolina State University, Biological and Agricultural Engineering Department, North Carolina Agricultural Extension Service, "Annotated Bibliography for the Project 'Rural Nonpoint Source Control Water Quality Evaluation and Technical Assistance,' " n.d.

2. International Joint Commission, International Reference Group on Great Lakes Pollution from Land Use Activities, *Evaluation of Remedial Measures to Control Non-point Sources of Water Pollution in the Great Lakes Basin* (Don Mills, Ontario: Marshall Macklin Monaghan, 1977), pp. 24, 25.

3. John A. Miranowski and Ruth Larson Bender, "Impact of Erosion Control Policies on Wildlife Habitat on Private Lands," *Journal of Soil and Water Conservation* 37, no. 5 (1982):288-91; Michael Duffy and Michal Hanthorn, *Returns to Corn and Soybean Tillage Practices,* prepared for U.S. Department of Agriculture, Agricultural Economic Report no. 508 (Washington, D.C.: U.S. Government Print-

ing Office, 1984); B. W. Menzel, "Agricultural Management Practices and the Integrity of Instream Biological Habitat," in Frank W. Schaller and George W. Bailey, eds., *Agricultural Management and Water Quality* (Ames, Iowa: Iowa State University Press, 1983), pp. 305-29; and D. H. Mueller, T. C. Daniel, and R. C. Wendt, "Conservation Tillage: Best Management Practice for Nonpoint Runoff," in *Environmental Management* 5, no. 1 (1981):41.

4. Erick E. Smith et al., "Cost-Effectiveness of Soil and Water Conservation Practices for Improvement of Water Quality," Douglas A. Haith and Raymond C. Loehr, eds., *Effectiveness of Soil and Water Conservation Practices for Pollution Control*, prepared for U.S. Environmental Protection Agency, Office of Research and Development, Environmental Research Laboratory, EPA 600/3-79-106 (Washington, D.C.: U.S. Government Printing Office, 1979); Gary L. Oberts, "Water Resources Management: Nonpoint Source Pollution Technical Report," U.S. Environmental Protection Agency, Region V, Chicago, Ill. (Springfield, Va.: National Technical Information Service, May 1982); International Joint Commission, *Evaluation of Remedial Measures to Control Non-point Sources of Water Pollution in the Great Lakes Basin*; M. T. Lee et al., "Economic Analysis of Erosion and Sedimentation: Hambaugh-Martin Watershed," prepared for University of Illinois at Urbana-Champaign, Illinois Agricultural Experiment Station, and State of Illinois, Institute for Environmental Quality, July 1974; and J. L. Baker and H. P. Johnson, "Evaluating the Effectiveness of BMPs from Field Studies," in Schaller and Bailey, eds., *Agricultural Management and Water Quality.*

5. Smith et al., "Cost-Effectiveness of Soil and Water Conservation Practices for Improvement of Water Quality," pp. 164, 173.

6. Sandra S. Batie, *Soil Erosion: Crisis in America's Croplands?* (Washington, D.C.: The Conservation Foundation, 1983), pp. 62-64.

7. Haith and Loehr, eds., *Effectiveness of Soil and Water Conservation Practices for Pollution Control*, pp. 407, 410.

8. Oberts, "Water Resources Management," p. 106; International Joint Commission, *Evaluation of Remedial Measures to Control Non-point Sources of Water Pollution in the Great Lakes Basin*, p. 19(b); Lee et al., "Economic Analysis of Erosion and Sedimentation," p. 12; and Baker and Johnson, "Evaluating the Effectiveness of BMPs from Field Studies," p. 292.

9. U.S. Department of Agriculture, Agricultural Research Service, and U.S. Environmental Protection Agency, Office of Research and Development, *Control of Water Pollution from Cropland*, vol. 1, *A Manual for Guideline Development* (Washington, D.C.: U.S. Government Printing Office, 1975), p. 63; International Joint Commission, *Evaluation of Remedial Measures to Control Non-point Sources of Water Pollution in the Great Lakes Basin*, pp. 19, 35.

10. U.S. Department of Agriculture and U.S. Environmental Protection Agency, *Control of Water Pollution from Cropland,* vol. 1, *A Manual for Guideline Development*, pp. 63, 71.

11. Douglas A. Haith, "Effects of Soil and Water Conservation Practices on Edge-of-Field Nutrient Losses," in Haith and Loehr, eds., *Effectiveness of Soil and Water Conservation Practices for Pollution Control*, p. 88.

12. Ibid., pp. 407, 410.

13. U.S. Department of Agriculture and U.S. Environmental Protection Agency, *Control of Water Pollution from Cropland,* vol. 1, *A Manual for Guideline Development*, p. 75.

14. Conservation Tillage Information Center, *1982 National Survey Conservation Tillage Practices* (Fort Wayne, Ind.: Conservation Tillage Information Center, 1983), p. 5.

15. Batie, *Soil Erosion*, p.. 66.

16. Maureen K. Hinkle, "Problems with Conservation Tillage, *Journal of Soil and Water Conservation* 38, no. 3 (1983):201-6; and John A. Miranowski, "Agricultural Impacts on Environmental Quality," in Ted L. Napier et al., eds., *Water Resources Research: Problems and Potentials for Agriculture and Rural Communities* (Ankeny, Iowa: Soil Conservation Society of America, 1983), p. 129.

17. U.S. Environmental Protection Agency, Office of Water Program Operations, Water Planning Division, *Nonpoint Source Pollution in the U.S.,* Report to the Congress (Washington, D.C.: U.S. Environmental Protection Agency, 1984), table A-1.

18. International Joint Commission, *Evaluation of Remedial Measures to Control Non-point Sources of Water Pollution in the Great Lakes Basin*, p. 33.

19. Lee A. Christensen and Patricia E. Norris, "A Comparison of Tillage Systems for Reducing Soil Erosion and Water Pollution," Agricultural Economic Report no. 499 (Washington, D.C.: U.S. Department of Agriculture, 1983), p. 15; Harold R. Cosper, Merlin W. Erickson, and Herbert Hoover, "Assessing the Potential for Conservation Tillage: A Case Study in the Maple Creek Watershed," NRE Staff Report (Washington, D.C.: U.S. Department of Agriculture, 1983), p. 4; and Jesse R. Russell and Lee A. Christensen, "Use and Cost of Soil Conservation and Water Quality Practices in the Southeast" (Washington, D.C.: U.S. Department of Agriculture, 1984).

20. Smith et al., "Cost-Effectiveness of Soil and Water Conservation Practices for Improvement of Water Quality," p. 172; and Christensen and Norris, "A Comparison of Tillage Systems for Reducing Soil Erosion and Water Pollution," p. 18.

21. Smith et al., "Cost-Effectiveness of Soil and Water Conservation Practices for Improvement of Water Quality," p. 165.

22. *Farm Agricultural Resources Management Conference on Conservation Tillage* (Ames, Iowa: Iowa State University, Cooperative Extension Service, 1982), p. 33; International Joint Commission, *Evaluation of Remedial Measures to Control Non-point Sources of Water Pollution in the Great Lakes Basin*; R. M. Cruse et al., "Tillage Effects on Corn and Soybean Production in Farmer-Managed, University-Monitored Field Plots," *Journal of Soil and Water Conservation* 38, no. 6 (1983):512-14; and Don Thill, "Reduced Tillage Weed Control," in *Conservation Tillage Conference Proceedings* (Moscow, Idaho: Soil Conservation Society of America, 1984), p. 40.

23. Haith and Loehr, eds., *Effectiveness of Soil and Water Conservation Practices for Pollution Control*, p. 377.

24. International Joint Commission, *Evaluation of Remedial Measures to Control Non-point Sources of Water Pollution in the Great Lakes Basin*, pp. 23, 33.

25. *Farm Agricultural Resources Management Conference on Conservation Tillage*; and J. W. Doran, "Microbial Changes Associated with Residue Management with Reduced Tillage," *Soil Science Society of America Journal* 44 (1980):518-24.

26. Smith et al., "Cost-Effectiveness of Soil and Water Conservation Practices for Improvement of Water Quality," p. 157.

27. Christensen and Norris, "A Comparison of Tillage Systems for Reducing Soil Erosion and Water Pollution"; James B. Atkins, "Pollution Control and Production Efficiency," *Extension Review*, Spring 1983, p. 47; and Garfield J. House et al., "Nitrogen Cycling in Conventional and No-Tillage Agroecosystems in the Southern Piedmont," *Journal of Soil and Water Conservation* 39, no. 3 (1984):194-200.

28. Smith et al., "Cost-Effectiveness of Soil and Water Conservation Practices for Improvement of Water Quality," p. 165.

29. U.S. Department of Agriculture and U.S. Environmental Protection Agency, *Control of Water Pollution from Cropland*, vol. 1, *A Manual for Guideline Development*, p. 71; H. P. Johnson and J. L. Baker, "Evaluating the Effectiveness of BMPs from Field Studies," in Schaller and Bailey, eds., *Agricultural Management and Water Quality*, chapter 15; and John C. Reardon, Lowell D. Hanson, and John Randolph, "Using EPA's Computerized Data Base (STORET) to Analyze for Agricultural Water Pollution," *Journal of Environmental Quality* 11, no. 3 (1982):430.

30. Baker and Johnson, "Evaluating the Effectiveness of BMPs from Field Studies," p. 288.

31. Mueller, Daniel, and Wendt, "Conservation Tillage," pp. 42, 45; Walter J. Rawls and H. H. Richardson, "Runoff Curve Numbers for Conservation Tillage," *Journal of Soil and Water Conservation* 38, no. 6 (1983):494-96; U.S. Environmental Protection Agency, *Nonpoint Source Pollution in the U.S.*, table A-1; J. L. Baker and J. M. Laflen, "Water Quality Consequences of Conservation Tillage," *Journal of Soil and Water Conservation* 38, no. 3 (1983):190-91; and J. L. Baker and H. P. Johnson, "The Effect of Tillage Systems on Pesticides in Runoff from Small Watersheds, *Transactions of the American Society of Agricultural Engineers* 22, no. 3 (1979):558.

32. Mueller, Daniel, and Wendt, "Conservation Tillage," p. 45; Oberts, "Water Resources Management," p. 106; and Baker and Laflen, "Water Quality Consequences of Conservation Tillage," pp. 190-91.

33. Oberts, "Water Resources Management," p. 106; and Haith, "Effects of Soil and Water Conservation Practices on Edge-of-Field Nutrient Losses," p. 88.

34. G. B. Triplett, Jr., B. J. Conner, and W. M. Edwards, "Transport of Atrazine and Simazine in Runoff from Conventional and No-Tillage Corn," *Journal of Environmental Quality* 7, no. 1 (1978):78.

35. U.S. Environmental Protection Agency, *Nonpoint Source Pollution in the U.S.*, table A-1.

36. Baker and Johnson, "The Effect of Tillage Systems on Pesticides in Runoff from Small Watersheds," p. 558.

37. Ibid.

38. Mueller, Daniel, and Wendt, "Conservation Tillage," p. 47; and Haith, "Effects of Soil and Water Conservation Practices on Edge-of-Field Nutrient Losses," p. 104.

39. Duffy and Hanthorn, "Returns to Corn and Soybean Tillage Practices," p. 3.

40. Smith et al., "Cost-Effectiveness of Soil and Water Conservation Practices for Improvement of Water Quality," p. 162.

41. Batie, *Soil Erosion*, p. 57.

42. U.S. Department of Agriculture and U.S. Environmental Protection Agency, *Control of Water Pollution from Cropland*, vol. 1, *A Manual for Guideline Development*, p. 79.

43. W. Luckmann, "Insect Control in Corn: Practices and Prospects" (Paper delivered at Pest Control Strategies: Understanding and Action, conference held at Cornell University, Ithaca, New York, 1977), cited in Smith et al., "Cost-Effectiveness of Soil and Water Conservation Practices for Improvement of Water Quality," p. 159.

44. Smith et al., "Cost-Effectiveness of Soil and Water Conservation Practices for Improvement of Water Quality," p. 165.

45. U.S. Department of Agriculture and U.S. Environmental Protection Agency, *Control of Water Pollution from Cropland*, vol. 1, *A Manual for Guideline Development*, p. 63.

46. Smith et al., "Cost-Effectiveness of Soil and Water Conservation Practices for Improvement of Water Quality," p. 172.

47. Ibid., p. 88; and U.S. Department of Agriculture, Soil Conservation Service, *Water Quality Field Guide* (Washington, D.C.: U.S. Government Printing Office, 1983), p. 35.

48. U.S. Department of Agriculture and U.S. Environmental Protection Agency, *Control of Water Pollution from Cropland*, vol. 1, *A Manual for Guideline Development*, p. 65.

49. M. F. Miller, *Cropping Systems in Relation to Erosion Control* (Columbia, Mo.: Missouri Agricultural Experiment Station, 1936), cited in Batie, *Soil Erosion*, p. 57.

50. U.S. Department of Agriculture and U.S. Environmental Protection Agency, *Control of Water Pollution from Cropland*, vol. 1, *A Manual for Guideline Development*, p. 86.

51. U.S. Comptroller General, *Agriculture's Soil Conservation Programs Miss Full Potential in the Fight Against Soil Erosion*, Report to the Congress, GAO/RCED-84-48 (Washington, D.C.: U.S. General Accounting Office, November 28, 1983), p. 52.

52. U.S. Department of Agriculture and U.S. Environmental Protection Agency, *Control of Water Pollution from Cropland*, vol. 1, *A Manual for Guideline Development*, p. 63; and International Joint Commission, *Evaluation of Remedial Measures to Control Non-point Sources of Water Pollution in the Great Lakes Basin*, p. 35.

53. North Carolina State University, Biological and Agricultural Engineering Department, North Carolina Agricultural Extension Service, "Best Management Practices for Agricultural Nonpoint Source Control: III. Sediment," Raleigh, N.C., August 1982, p. 9.

54. U.S. Department of Agriculture and U.S. Environmental Protection Agency, *Control of Water Pollution from Cropland*, vol. 1, *A Manual for Guideline Development*, p. 83; and International Joint Commission, *Evaluation of Remedial Measures to Control Non-point Sources of Water Pollution in the Great Lakes Basin*, p. 35.

55. Haith and Loehr, eds., *Effectiveness of Soil and Water Conservation Practices for Pollution Control*, p. 8.

56. Smith et al., "Cost-Effectiveness of Soil and Water Conservation Practices for Improvement of Water Quality," p. 166.

57. International Joint Commission, *Evaluation of Remedial Measures to Control Non-point Sources of Water Pollution in the Great Lakes Basin*, p. 31b.

58. Ibid.

59. Tennessee Department of Public Health, Division of Water Quality Control, "Water Quality Management Plan for Agriculture in Tennessee," November 1978, p. III-7; and G. B. White and E. J. Partenheimer, "The Economic Implications of Erosion and Sedimentation Control Plans for Selected Pennsylvania Dairy Farms," in Raymond C. Loehr et al., eds., *Best Management Practices for Agriculture and Silvaculture* (Proceedings of the 1978 Cornell Agricultural Waste Management Conference) (Ann Arbor, Mich.: Ann Arbor Science Publishers, 1979), p. 355.

60. Christine A. Shoemaker and Marion O. Harris, "The Effectiveness of SWCPs in Comparison with Other Methods for Reducing Pesticide Pollution," in Haith and Loehr, eds., *Evaluation of Soil and Water Conservation Practices for Pollution Control,* p. 282; Oberts, "Water Resources Management," p. 107; Menzel, "Agricultural Management Practices and the Integrity of Instream Biological Habitat," p. 310; North Carolina State University, "Best Management Practices for Agricultural Nonpoint Source Control," p. 28. U.S. Department of Agriculture and U.S. Environmental Protection Agency, *Control of Water Pollution from Cropland,* vol. 1, *A Manual for Guideline Development*," p. 68; and U.S. Environmental Protection Agency, *Nonpoint Source Pollution in the U.S.*., table A-1.

61. Haith, "Effects of Soil and Water Conservation Practices on Edge-of-Field Nutrient Losses," p. 88.; and Menzel; "Agricultural Management Practices and the Integrity of Instream Biological Habitat," p. 310.

62. Oberts, "Water Resources Management," p. 107.

63. International Joint Commission, *Evaluation of Remedial Measures to Control Non-point Sources of Water Pollution in the Great Lakes Basin*, p. 22.

64. Ibid.

65. American Farmland Trust, *Soil Conservation in America: What Do We Have to Lose?* (Washington, D.C.: American Farmland Trust, 1984), pp. 40-41.

66. Ibid.

67. Mueller, Daniel, and Wendt, "Conservation Tillage," p. 33; and U.S. Department of Agriculture and U.S. Environmental Protection Agency, *Control of Water Pollution from Cropland*, vol. 1, *A Manual for Guideline Development*, p. 63.

68. C. Arden Pope, Shashanka Buide, and Earl O. Heady, "The Economics of Soil and Water Conservation Practices in Iowa: Results and Discussion," CARD Report no. 109, SWCP Series II (Ames, Iowa: Iowa State University, Center for Agricultural and Rural Development, February 1983, p. 23; Smith et al., "Cost-Effectiveness of Soil and Water Conservation Practices for Improvement of Water Quality," p. 164; International Joint Commission, *Evaluation of Remedial Measures to Control Non-point Sources of Water Pollution in the Great Lakes Basin*, p. 22b.

69. U.S. Environmental Protection Agency, *Nonpoint Source Pollution in the U.S.*, table A-1.

70. Tammo S. Steenhuis and Michael F. Walter, "Definitions and Qualitative Evaluation of Soil and Water Conservation Practices," in Haith and Loehr, eds., *Effectiveness of Soil and Water Conservation Practices for Pollution Control*, p. 22.

71. Pope, Buide, and Heady, "The Economics of Soil and Water Conservation Practices in Iowa," p. 23.

72. International Joint Commission, *Evaluation of Remedial Measures to Con-*

trol Non-point Sources of Water Pollution in the Great Lakes Basin, p. 94; and Dane County Regional Planning Commission, *Appendix E—Agricultural Nonpoint Source Analysis* (Madison, Wisc.: Dane County Regional Planning Commission, 1980), p. E-66.

73. Smith et al., "Cost-Effectiveness of Soil and Water Conservation Practices for Improvement of Water Quality," p. 166.

74. International Joint Commission, *Evaluation of Remedial Measures to Control Non-point Sources of Water Pollution in the Great Lakes Basin*, p. 81.

75. Ibid., p. 81b.

76. Oberts, "Water Resources Management," p. 107.

77. Ibid.

78. Ibid., pp. 106-9.

79. Menzel, "Agricultural Management Practices and the Integrity of Instream Biological Habitat," p. 325.

80. International Joint Commission, *Evaluation of Remedial Measures to Control Non-point Sources of Water Pollution in the Great Lakes Basin*, p. 93.

81. Ibid., pp. 39, 93.

82. Menzel, "Agricultural Management Practices and the Integrity of Instream Biological Habiatat," p. 310.

83. International Joint Commission, *Evaluation of Remedial Measures to Control Non-point Sources of Water Pollution in the Great Lakes Basin*, p. 93.

84. Oberts, "Water Resources Management," p. 107.

85. North Carolina State University, "Best Management Practices for Agricultural Nonpoint Source Control," pp. 9-10.

86. Menzel, "Agricultural Management Practices and the Integrity of Instream Biological Habitat," p. 310.

87. Oberts, "Water Resources Management," p. 107.

88. Triplett et al., "Transport of Atrazine and Simazine in Runoff from Conventional and No-Tillage Corn," p. 78.

89. International Joint Commission, *Evaluation of Remedial Measures to Control Non-point Sources of Water Pollution in the Great Lakes Basin*, p. 80b.

90. Oberts, "Water Resources Management," p. 106; and Menzel, "Agricultural Management Practices and the Integrity of Instream Biological Habiatat," p. 310.

91. Oberts, "Water Resources Management," p. 106; and Menzel, "Agricultural Management Practices and the Integrity of Instream Biological Habitat," p. 310.

92. Michael F. Walter, Tammo S. Steenhuis, and Hanneke P. DeLancey, "The Effects of Soil and Water Conservation Practices on Sediment," in Haith and Loehr, eds., *Effectiveness of Soil and Water Conservation Practices for Pollution Control*, p. 64.

93. International Joint Commission, *Evaluation of Remedial Measures to Control Non-point Sources of Water Pollution in the Great Lakes Basin*, p. 16.

94. Oberts, "Water Resources Management," p. 107.

95. Tennessee Department of Public Health, "Water Quality Management Plan for Agriculture in Tennessee," p. III-17.

96. U.S. Department of Agriculture, "Speeches and Major Press Releases," April 20-April 27, 1984, p. 2.

97. P. D. Robillard and M. F. Walter, "A Framework for Selecting Agricultural Nonpoint Source Controls," in Schaller and Bailey, *Agricultural Management and Water Quality*, p. 335.

98. A. S. Donigian et al., *Agricultural Runoff Management (ARM) Model, Version II: Refinement and Testing*, EPA 600/3-77-098 (Washington, D.C.: U.S. Environmental Protection Agency, 1977); cited in Baker and Laflen, "Water Quality Consequences of Conservation Tillage," p. 188.

99. D. R. Timmons, R. E. Burwell, and R. F. Holt, "Nitrogen and Phosphorus Losses in Surface Runoff from Agricultural Land as Influences by Placement of Broadcast Fertilizers," *Water Resources Research* 9 (1973):658-67; cited in Baker and Laflen, "Water Quality Consequences of Conservation Tillage," p. 188.

100. International Joint Commission, *Evaluation of Remedial Measures to Control Non-point Sources of Water Pollution in the Great Lakes Basin*, p. 27.

101. See, generally, *Conservation Tillage Conference Proceedings.*

102. J. C. Baker and J. M. Laflen, "Runoff Losses of Surface-Applied Herbicides as Affected by Wheel Tracks and Incorporation," *Journal of Environmental Quality* 8 (1979):602-7; cited in Baker and Laflen, "Water Quality Consequences of Conservation Tillage," p. 188.

103. International Joint Commission, *Evaluation of Remedial Measures to Control Non-point Sources of Water Pollution in the Great Lakes Basin*, p. 28.

104. Oberts, "Water Resources Management," p. 107.

105. International Joint Commission, *Evaluation of Remedial Measures to Control Non-point Sources of Water Pollution in the Great Lakes Basin*, p. 24.

106. Shoemaker and Harris, "The Effectiveness of SWCPs in Comparison with Other Methods for Reducing Pesticide Pollution," p. 230.

107. David Bull, *A Growing Problem: Pesticides and the Third World Poor* (Oxford, England: OXFAM, 1982), pp. 126-28.

108. International Joint Commission, *Evaluation of Remedial Measures to Control Non-point Sources of Water Pollution in the Great Lakes Basin*, pp. 25, 27.

109. Burton C. English et al., eds., *RCA Symposium: Future Agricultural Technology and Resource Conservation*, keynote address (Ames, Iowa: Iowa State University, Center for Agricultural and Rural Development, 1983), p. 23.

110. U.S. Environmental Protection Agency, *Nonpoint Source Pollution in the U.S.*, p. A-1.

111. Shoemaker and Harris, "The Effectiveness of SWCPs in Comparison with other Methods for Reducing Pesticide Pollution," p. 235.

112. D. W. Fitzsimmons et al., "Evaluation of Measures for Controlling Sediment and Nutrient Losses from Irrigated Areas," prepared for U.S. Environmental Protection Agency, Robert M. S. Kerr Environmental Research Laboratory, Ada, Oklahoma, EPA-600/2-78/138, 1978, p. 7.

113. Ibid.

114. Ibid., p. 3.

115. Ibid.

116. U.S. Department of Agriculture, Soil Conservation Service, Patterson Field Office and River Basin Planning Staff, "Comparison of Alternative Management Practices Mini-Report: Spanish Grant Drainage District and Crow Creek Pilot Study Areas, Stanislaus County, California," Patterson Field Office, Patterson, California, and River Basin Planning Staff, Davis, California, November 1979, p. 12.

117. Fitzsimmons et al., "Evaluation of Measures for Controlling Sediment and Nutrient Losses from Irrigated Areas," p. 3.

118. Ibid., pp. 3, 4.

119. Harold G. Halcrow and Wesley D. Seitz, "Soil Conservation Policies, In-

stitutions and Incentives: A Summary,'' in Max Schnepf, ed., *Natural Resources Policy: Research Strategies for the Future* (proceedings of a conference) (Ankeny, Iowa: Soil Conservation Society of America, 1982), p. 7.

120. Smith et al., "The Cost-Effectiveness of Soil and Water Conservation Practices for Improvement of Water Quality,'' p. 165.

121. Ibid., p. 168.

122. U.S. Department of Agriculture, Soil Conservation Service, Forest Service, and Economics, Statistics, and Cooperatives Service, *Palouse Cooperative River Basin Study* (Washington, D.C.: U.S. Government Printing Office, 1979), pp. 91, 104.

123. Haith, "Effects of Soil and Water Conservation Practices on Edge-of-Field Nutrient Losses,'' p. 88.

124. James L. Arts and William L. Church, "Soil Erosion—The Next Crisis?" *Wisconsin Law Review* 1982, no. 4 (1982):571.

125. Arts and Church, "Soil Erosion,'' p. 571.

126. U.S. Department of Agriculture, *Comparison of Alternative Management Practices,* p. 10.

127. Robillard and Walter, "A Framework for Selecting Agricultural Nonpoint Source Controls," p. 337.

128. Ibid.

129. C. Robert Taylor, Klaus K. Frohberg, and Wesley D. Seitz, "Potential Erosion and Fertilizer Controls in the Corn Belt: An Economic Analysis," *Journal of Soil and Water Conservation* 33, no. 4 (1978); J. C. Wade and E. O. Heady, "Measurement of Sediment Control Impacts on Agriculture,'' *Water Resources Research* 14, no. 1 (1978); and Iowa State University, "Economic Impacts of Water Pollution Control Act of 1972: Irrigated and Non-Irrigated Agriculture,'' prepared for the National Commission on Water Quality, July 1975.

130. Smith et al., "Cost-Effectiveness of Soil and Water Conservation Practices for Improvement of Water Quality,'' p. 196.

131. Clayton W. Ogg, Arnold B. Miller, and Kenneth C. Clayton, "Agricultural Program Integration to Achieve Soil Conservation" no date.

132. Grace S. Brush, "Stratigraphic Evidence of Eutrophiciation in an Estuary,'' *Water Resources Research* 20, no. 5 (1984):531-41.

133. U.S. Environmental Protection Agency, *Chesapeake Bay: A Framework for Action* (Washington, D.C.: U.S. Government Printing Office, 1983).

134. Ibid., pp. 156-57.

135. John A. Miranowski and Ruth Larson Bender, "Impact of Erosion Control Policies on Wildlife Habitat on Private Lands.''

7. Options for Government

Few federal programs have involved so many people and so much money over so long a period of time as have the nation's attempts to control soil erosion. Over the past 40 years, an estimated $15 to $30 billion has been spent for soil and water conservation.[1] These programs have established a large body of information on soil conservation measures and an extensive network for providing technical and financial assistance, both of which must form the basis of any efforts to control off-farm impacts.

However, erosion rates remain high. Since U.S. soil conservation programs were begun, they have sought primarily to reduce erosion-caused losses in farmland productivity. Relatively little attention has been given to reducing erosion's off-farm impacts. Soil conservation programs need to incorporate these off-farm concerns with existing on-farm goals, although any program that focuses only on off-farm effects probably will be just as inefficient as past programs that addressed only on-farm problems.*

Developing a strategy that both effectively addresses erosion's off-farm impacts and reflects budgetary and political realities will be no easy task. It will require effective policy tools to encourage landowners to adopt efficient erosion-control techniques, careful targeting of control

---

*Nevertheless, a philosophical case can be made that government should get involved in controlling soil erosion on privately owned land only when it does so because of the off-farm impacts. One principle of American politics is the idea that government programs are most appropriate (or, according to some, only appropriate) when a person making a decision does not experience all the benefits or costs associated with that decision. With soil erosion, a landowner generally experiences the costs caused by reduced productivity but not those associated with off-farm impacts. Thus, according to this principle, any government involvement in reducing soil erosion on private farmland should focus on the off-farm impacts, not the on-farm productivity losses.

efforts to those lands that are the major sources of the pollutants causing the problems, and a sensitive implementation strategy that effectively coordinates efforts at the federal, state, and local levels.

POLICY TOOLS

Past soil conservation efforts in the United States have relied predominantly on education, technical assistance, and cost sharing to stimulate the adoption of conservation measures by individual landowners. It is increasingly a matter of debate whether these policy tools, by themselves, can control agricultural soil erosion. Fortunately, other policy tools also are available to the federal government.

Education and Technical Assistance

After 50 years of federal efforts to promote soil conservation, most people are aware of the problem of soil erosion and familiar with ways of dealing with it. However, several different surveys of farmers have indicated that further education still is needed.[2] The more farmers understand and believe in the goals of conservation programs, the easier the programs will be to implement.[3]

Sometimes, farmers may not realize when erosion is occurring on their land, because it can occur very slowly and subtly even if many tons of soil are being carried off. In addition, if farmers do understand how much erosion is occurring, they may not be aware of the problems that the erosion is causing, particularly after it leaves their farms. Farmers who do not see erosion significantly affecting their own land's productivity often may see no reason to try to control the erosion.

Even farmers who do know that they need to control erosion on their land may not be sure how to do so most effectively and cost-efficiently. Farmers can be understandably reluctant to invest in costly systems that they perceive as ineffective.

Cost Sharing

Another tool that has traditionally been used to stimulate the adoption of soil conservation measures is cost sharing. In the past, this approach has been purely voluntary, with the available funds spread among those farmers who requested them first.[4] Depending on the erosion-control techniques that individual farmers proposed to adopt, the federal government has paid up to 75 percent of a project's initial cost.

Unfortunately, with this purely voluntary approach, much money has been spent on lands that were not eroding seriously.[5] Apparently, much of this money was used for productivity improvements in general rather than for erosion control in particular. USDA currently is experimenting with dealing with this problem by having the level of government cost sharing vary according to the amount of soil loss that a conservation project could prevent.[6]

Many cost-sharing programs apply only to initial investments and not to operation and maintenance costs, thus providing incentives favoring capital investment over management techniques that may be cheaper. Such programs may also fail to provide incentives to maintain investments properly if farmers do not perceive them to be in their own best interest. In some cases, USDA has attempted to deal with this problem by entering into contracts with landowners and requiring them to repay part of their initial subsidies if they do not maintain the erosion control for a specified period.

Other Economic Incentives

Cost sharing may not be the most effective incentive for stimulating the adoption of soil conservation measures. Several other incentives have been suggested to either financially penalize farmers for not practicing appropriate erosion-control measures or, alternatively, reward them if they do.

One penalty incentive is a soil-loss tax. This system is sometimes proposed as the most efficient way to stimulate the adoption of appropriate levels of soil conservation measures.[7] Under such a scheme, farmers would be charged on the basis of the amount of soil (and presumably associated contaminants) carried off their land. The tax would be set at a level that represented the costs of the targeted erosion to society, both in terms of its off-site impacts and the long-term losses in productivity.

In its purest form, the size of such a tax probably would vary between areas of the country, and even between different farms in the same area, to reflect the differences in the magnitude of the off-site impacts created by the erosion as well as the differences in productivity losses. However, determining the efficient level of the tax would be very difficult,* and monitoring the amounts of soil and con-

*Theoretically, it should be no more difficult to determine the efficient tax level than to identify the efficient control technique, since that technique would be the one resulting in the amount of control that would be produced by a tax.

taminants being carried off an individual landowner's property could be even more formidable.

Finally, such a scheme would have significant implications for agricultural income. One analysis found that, under a soil-loss tax scheme, farmers would end up paying substantially more in taxes than to control erosion.[8] Farmers would see relatively little improvement in efficiency but they would feel significant financial impacts. These considerations, along with serious difficulties that would be involved in implementing such a scheme, argue strongly against the soil-loss tax proposal.

Cross-compliance, another type of penalty program, appears more feasible, although it has been the subject of heated debate since it was originally proposed in the late 1970s.[9] Under a cross-compliance scheme, the benefits of agricultural income-support and subsidy programs would be withheld from farmers who failed to practice approved soil conservation measures. The types of programs that might be involved in such a scheme include federal price-support programs, subsidized loan programs, and state and local preferential taxation programs.[10]

The argument in favor of cross-compliance is that society should not subsidize farmers who are unwilling to care for their lands in a socially desirable matter. An argument against it, however, is that the government's other income-maintenance and subsidy programs have been adopted for their own particular purposes and should not be linked to soil conservation unless the linked program actually contributes to increased soil erosion.

The effectiveness of a cross-compliance approach would depend on three major factors—the extent to which the other programs included were used by farmers with serious erosion problems; the extent to which the threat to withhold the other programs would provide sufficient incentive for farmers to take action; and the extent to which farmers could effectively control their problems.

The evidence on the first of these factors is that many income-support programs are not benefiting farmers with serious erosion problems. One analyst concluded, for instance, that "most of the worst erosion does not occur on farms that participate regularly in commodity programs."[11]

There is little evidence on the second factor, except for the obvious fact that cross-compliance would provide little incentive in situations where it did not apply. In addition, this incentive would decrease substantially during agricultural boom times when price support programs were unneeded and largely unused. It is during those times that the amount of land in crops generally increases, so that soil erosion may be at its most serious. One analysis has predicted a 70 percent increase in soil erosion if crop prices regain the heights (adjusted for inflation) they achieved in the early 1970s.[12]

Substantial debate exists around the third factor. A scheme that required farmers to reduce erosion rates to a specified level before they could benefit from other programs might well keep the most erosive land from being included, since it is almost impossible to achieve an acceptable erosion rate on such land without taking it out of row-crop production entirely.[13] In addition, low-income farmers who depend on loans from the Farmers Home Administration could be unable to afford the expenditures required.

Penalty proposals such as soil-loss taxes and cross-compliance are apt to be unpopular among many in the farming community and, therefore, in Congress.[14] Nevertheless, some opinion surveys are showing more and more farmers to be open to them, with a majority in some states even appearing to favor some form of cross-compliance.[15]

Understandably, however, farmers tend to prefer positive economic incentives such as cost sharing. Other such incentives that have been suggested include breaks on either income or property taxes, increases in the level of benefits provided by other programs if conservation is practiced (a positive cross-compliance program), special benefit programs for farmers who adopt soil conservation techniques, and straight economic bonuses for reducing erosion (the opposite of the soil-loss tax).[16] The variable cost-sharing program being tested by USDA is a type of positive economic incentive, similar to an erosion-reduction bonus.

Although positive incentives generally are more popular among farmers, some have the same implementation problems as the negative incentives and all imply increased budget outlays (or reduced tax revenues) at the federal, state, and local levels—not a popular suggestion during periods of budget austerity.

Regulatory Proposals

The biggest question confronting the soil conservation community is whether erosion can be controlled without the establishment of some regulatory authority. Regulation could take any of several forms. It could place limits on the amount of sediment and other contaminants leaving a farm or require that specific soil conservation measures be adopted. Alternatively, zoning could be used to restrict lands to specific uses or cropping schemes unless specified soil erosion measures were adopted.[17] Although many dismiss regulatory schemes as unlikely ever to be adopted to promote soil conservation,[18] they are not completely unknown in agriculture. Programs limiting the production of milk, tobacco, and other commodities are widely accepted.[19]*

Moreover, there is little disagreement that an adequate legal basis does exist for establishing regulatory programs.[20] Twenty-six states already have, in their laws establishing conservation districts, provisions allowing those districts to enact limited land-use ordinances.[21] In addition, the model state act to control soil erosion (developed by the National Association of Conservation Districts and the Council of State Legislators) authorizes such enforcement measures as injunctions and penalties in extreme cases to force compliance with soil conservation requirements.[22] As of 1980, 16 of 22 soil-erosion-control statutes in states such as Iowa, Pennsylvania, Michigan, and South Dakota included regulatory components (figure 7.1).[23]

Three major questions surround regulatory approaches to controlling soil erosion: (*a*) are such programs necessary; (*b*) if they are necessary, can they be enacted; and (*c*) even if they are enacted, will they be implemented? Agreement that such programs may be necessary, at least in some cases, appears to be increasing.[24] In addition, the adoption of state and local statutes incorporating regulatory alternatives proves that such programs can be enacted, despite the unlikelihood that the federal government will adopt any regulatory schemes in the near future. And farmers are increasingly indicating some receptivity to stronger programs.[25]

This still leaves the question of whether such programs will be enforced if they are adopted.[26] Even such noncontroversial current requirements as the one associated with the payment-in-kind (PIK) income-support program—that the land taken out of production be planted in a cover crop—is likely not to be enforced. Surveys by the National Association of Conservation Districts found that almost a quarter of such lands were in danger of suffering excessive soil erosion because their cover crops were inadequate.[27] Leaving such lands barren could cause serious problems since the land taken out of production is often steep-sloped, highly erodible land.

Other Options

Two tools that do not fall neatly into any of the above categories are enforceable contracts and public purchase. As mentioned above, USDA has begun specifying in some long-term contracts that farmers who

*These programs, of course, have the purpose of raising prices and, therefore, are bound to be more popular with farmers than ones that impose costs.

Figure 7.1
State Soil-Erosion-Control Statues

	Type of program		Type of state law		
State	Regulatory	Non-regulatory	Erosion and sediment control	Conservation districts	Water quality and stream control
Delaware	X		X		
District of Columbia	X		X		
Georgia	X		X		
Hawaii		X	X		
Illinois		X		X	
Iowa	X			X	
Maine	X				X
Maryland	X		X		
Michigan	X		X		
Minnesota		X		X	
Montanta		1			X
Nevada					X
New Hampshire	X				X
New Jersey	X		X	X	
New York		X		X	
North Carolina	X	X			
Ohio	X	2	X	X	
Pennsylvania	X				3
South Carolina	X		X		
South Dakota	X		X		
Virginia	X		X	X	
Virgin Islands	X		X		

1. Diffuse-source control is a voluntary program within a regulatory program.
2. Animal-waste pollution-abatement program is regulatory.
3. Authorities are contained in laws and regulations.

Source: National Conference of State Legislatures, *State Soil Erosion and Sediment Control Laws*, November 1980.

receive cost-sharing funds must repay part of their subsidies if they do not properly maintain their conservation measures. Contracts also could call for nonperformance penalties or could allow a government agency to go to court to get an injunction mandating that a contract be obeyed. The more aggressive the terms of a contract, however, the less likely farmers will be to enter into one.

For some lands, the best approach may be outright public purchase, either of the property in fee simple or of an easement that would permanently keep the property from being planted with row crops.[28] On particularly erosive land, this approach could be much less costly than any alternative means of erosion control. Public purchase would be a permanent solution, not one that operates for only one year at a time. (The latter has been the case with many land-retirement programs implemented either to provide price supports or promote soil conservation.) However, purchasing land outright would make the government responsible for managing many dispersed small parcels. Purchasing easements would eliminate this problem.

Choosing Tools

Deciding which tools to use may be the most difficult part of developing erosion-control policies. Three basic criteria need to govern the choice. The first is efficiency. Selected policies should encourage farmers to adopt cost-effective control techniques on lands where they are needed. For instance, as mentioned above, cost sharing can lead landowners to adopt conservation measures other than the most cost-effective options and, like other voluntary tools, does not always induce measures to be taken where they are needed.

The second criterion is equity, meaning that similar cases should be treated in a similar manner. However, with agricultural policy in general, and soil erosion policy in particular, the concept of equity is often vague. The best one analyst could do was to define it as "the fair distribution of program costs and benefits among constituents."[29]

Traditional concepts of equity run strongly counter to the need for efficiency. Legislators and others may argue that moving away from the traditional voluntary approach to erosion control—for instance, by requiring that soil conservation measures be undertaken where there are particularly serious erosion problems—would be unfair. This attitude reflects both the currently frequent disregard for efficiency considerations and the curious notion of equity that apparently pervades agricultural programs—that it would be unfair to require one farmer with serious erosion problems to adopt control techniques to prevent off-farm impacts since a similar farmer without erosion difficulties would not also be forced to spend money on erosion control.

This concept of equity certainly is inconsistent with the heavy requirements forced on other contributors to water pollution. A program that required landowners to adopt pollution-control efforts, as unfair as it might appear when measured against past agricultural policies, would redress some of the inequities that traditionally have existed between the requirements governing point and nonpoint sources of pollution.

Nevertheless, these feelings about what is fair have significant practical and political importance. Any erosion-control program is likely to fail if the landowners who must implement it perceive it as inequitable and refuse to cooperate with it.

This relates to the third criterion, practicability. This concept is even vaguer than equity. Even if a scheme is efficient and equitable, it can still be impractical for political, institutional, budgetary, or technical reasons. In addition, the definition of what is practical can change substantially over time and from one place to another. Thus, any program should be sufficiently flexible to incorporate new elements in the future.

The tools that are most practical are those that are most easily accepted by the individuals creating the problems. Education and cost sharing are two such tools, but, as indicated above, it is very unlikely that these two tools by themselves can solve the entire erosion problem.[30] Some landowners, for economic or other reasons, are likely to remain recalcitrant, requiring more than just persuasion.

The government's *entire* soil conservation program will have to be perceived as fair and reasonable if it is to be accepted. One important aid in achieving that goal would be defining clearly the conditions under which the more aggressive tools in the program could be brought into play. For instance, such options might be statutorily restricted to situations where erosion rates are particularly high or where landowners have had substantial opportunity to question the legitimacy and equity of what they are being required to do.

Of course, just defining those conditions probably would not eliminate all opposition. Landowners will have to believe that the conditions are reasonable and will be strictly followed. Farmers often see giving government bureaucrats the authority to make decisions about how they should use their lands as a substantial threat, regardless of how reasonably that authority may be exercised.* Nevertheless, no matter how difficult the more aggressive tools may be to enact, a truly effective and efficient government program probably will require that they be available for use when needed.

TARGETING

No matter which tools are adopted, both the private and public sectors must spend significant amounts of money for a control program to be effective. The public sector's expenditures will need to include at least administrative, research, planning, technical assistance, and enforcement personnel and perhaps substantial financial resources for cost sharing and other incentive programs. The private sector will have to assume some or all of the costs of adopting and maintaining controls.

For those resources to be spent efficiently, they must be focused on the most serious causes of the problems being experienced. Most of the United States's 2.264 billion acres of land is eroding to some degree.[31] Over 400 million acres of that land (approximately 18 percent) is cropland, and almost one-fourth of that (95 million acres) is eroding at more than five tons per acre per year.[32] Immediately controlling erosion from all this land would be extremely inefficient, even if it were possible. As a result, conservation program managers must be able to identify which particular lands should be controlled first. This is the concept of targeting.

Targeting must occur at all levels of government. At the national level, funds need to be targeted to different states or regions of the country. Once this allocation has taken place, the funds must again be targeted to specific localities. At the local level, they must be targeted further to specific fields. Different information may be needed for each of these levels, substantially complicating the problem of developing an efficient targeting scheme. And, if the targeting that occurs at any of the different levels is inefficient, the efficiency of the entire program may be seriously compromised.

The concept of targeting is surprisingly controversial within the soil conservation community. Decisions on who gets what amounts of federal conservation funds traditionally have been based primarily on

---

*Some state statutes, such as Iowa's, have attempted to mitigate this threat by allowing action to be taken only after another farmer, who is being affected by the sediment, makes a formal complaint.

two criteria—landowners' receptivity and geographical equity.[33] As noted in the discussion of cost-sharing programs, available money generally has been spread among whichever farmers have applied for it first. However, when USDA's Agricultural Stabilization and Conservation Service investigated where those funds have gone, they found that some 52 percent of the assistance have been spent on lands eroding at less than five tons per acre per year.[34] Of the subsidies spent to encourage the growing of less erosive crops, 78 percent of the lands treated have been eroding at less than five tons per acre and 25 percent at less than one ton per acre.[35] Meanwhile, many of the nation's most erosive lands continue to have virtually no conservation measures.[36] Much of the federal assistance apparently has been used more to increase productivity than to decrease erosion.[37]

In reaction to the inefficiency of this allocation scheme, USDA's Soil Conservation Service (SCS) initiated a new targeting program in 1981, when $4 million, or 1.7 percent of the total "Conservation Technical Assistance" funds, were targeted to four specific geographical areas.[38] The SCS plans to expand this targeting program, giving it 25 percent of the total Conservation Technical Assistance budget by 1987.[39]*

The SCS also has begun restricting cost-sharing assistance to lands that are eroding at more than two tons per acre to prevent its resources from being used up on lands that are eroding relatively slowly.

None of these changes, however, explicitly considers the problem of off-farm impacts. Doing so would markedly complicate an erosion-control targeting scheme. While a program concerned only with on-farm impacts must be concerned primarily with how much erosion is occurring in different regions and on different fields, a targeting scheme focusing on off-farm effects must start with two different considerations. First, where are the off-farm impacts occurring, and how serious are they? Second, which areas and particular fields are the primary sources of the sediment and other pollutants causing these impacts?

There are several different approaches to targeting that could be adopted. Each would have its own advantages and disadvantages, and some would be more appropriate at particular levels of targeting or under certain types of conditions than would others. Some approaches would focus more on the location of impacts, and others on potential

---

*Not all the funds are targeted on the basis of soil erosion problems. Water conservation and other program purposes are also included.

sources of the problems. Any of these approaches would raise difficult conceptual problems. And currently there is inadequate information available to implement most of them with great precision.

Importance of Impacts

Deciding which erosion-related off-farm impacts are of most concern need not be a problem if they occur congruently in one limited area. For instance, the siltation and eutrophication of a lake may create recreation, aquatic wildlife, and water-storage and -usage problems at the same location and for the same reasons. In such cases, reducing one impact will reduce them all.

Unfortunately, however, such congruency is likely to be unusual, especially when a national targeting scheme is being attempted. The impacts that are most important on a national basis may not be the same as those that are important at the local level. For instance, downstream flooding might be considered very important nationally but relatively unimportant in upstream localities. In such cases, should local officials be allowed to allocate their funds to prevent local impacts, such as the deterioration of local fishing areas (which might do little to protect against the downstream flooding), or should they be forced to disregard the local impacts to help prevent the national problem?

One approach to evaluating the relative importance of the different impacts would be to consider the economic costs they impose. Unfortunately, many cost estimates (such as those contained in chapter 5) remain quite uncertain. Nor does it appear that some of those estimates could be substantially improved with any ease. As a result, such an approach would tend to bias expenditures in favor of those impacts for which economic costs can be identified and estimated relatively easily, not necessarily those that are the most serious.

Location of Impacts

The second problem with developing a national targeting scheme is the probability that the information necessary for its implementation exists more at the local level than at the national or regional level. Local people should be able easily to identify off-stream users of a river's water, areas prone to flooding, reservoirs and lakes fed by a waterway, and valuable recreation areas. They might even be able to identify the most valuable fish and wildlife habitats, although, even at the local level, some types of impacts would be difficult to identify. Stream aggradation, for instance, occurs so slowly that it may not even be noticed. People may know that flooding is more frequent and more serious than it was previously but have no idea whether the cause is more runoff coming from upstream, more people having built houses in the floodplains, less storage existing in the floodplain upstream, or the stream having aggraded.

At the national level, identifying the location impacts could be more difficult. Necessary information does not exist on many subjects and, even when it does, may not be in the form needed. For instance, although information does exist on national flood damages and on the location of navigation facilities across the United States, apparently no similar information is available on the locations of fish and wildlife habitats, recreational areas, off-stream users, or other types of impacted areas. What limited information is available on these subjects generally is not precisely what is needed to implement a targeting scheme efficiently. Although existing information may indicate the location of sites such as reservoirs, off-stream users, and fish spawning areas that theoretically could be affected by sediment and associated contaminants, those sites are not necessarily the ones that are actually suffering erosion-related damage. Similarly, merely knowing where flood damages are being experienced does not provide information about the relative importance of sediment and sedimentation in causing the damages. A city in New England may experience serious floods, but soil erosion may be a minor factor in those damages.

One substitute for information on the location of specific impacts could be data showing the presence in ambient water of erosion-related contaminants such as suspended solids, pesticides, nutrients, and dissolved solids. In fact, several states have adopted or proposed such an approach in preparing their own plans to control nonpoint-source pollution under section 208 of the Clean Water Act.[40] This approach, however, also creates a variety of problems. No one measure of erosion-related contamination exists; nor is there any accepted procedure for combining the several measures that do exist into one targeting scheme. Moreover, problems with one type of contaminant can be serious in areas where there are few problems with other contaminants (compare figures 7.2-7.5). Even if a targeting scheme were based on only one pollution problem, that problem would not be entirely solved, since there would still be a question whether the targeting should be aimed at the total amount of sediment being transported or the sediment concentration in the water. In some cases (for instance, sedimentation in reservoirs), the amount of sediment being transported probably would be the critical measure. In other cases, sediment concentration might be a more appropriate indicator.

Figure 7.2
Extent of Nonpoint Pollutants—
Suspended Solids

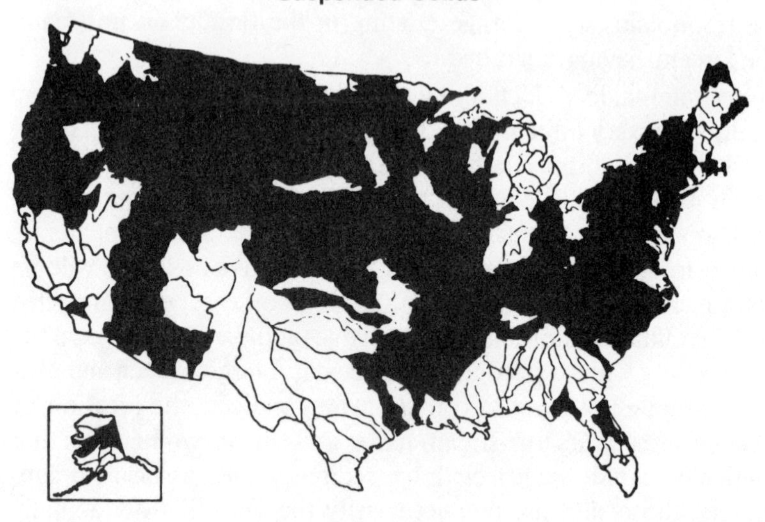

■ Represents basins affected by specific pollutant

Source: U.S. Department of Agriculture, Soil Conservation Service, "Environmental Impact Statement, Rural Clean Water Program" (Washington, D.C.: U.S. Department of Agriculture, 1978), p. 22.

Figure 7.3
Extent of Nonpoint Pollutants—
Pesticides

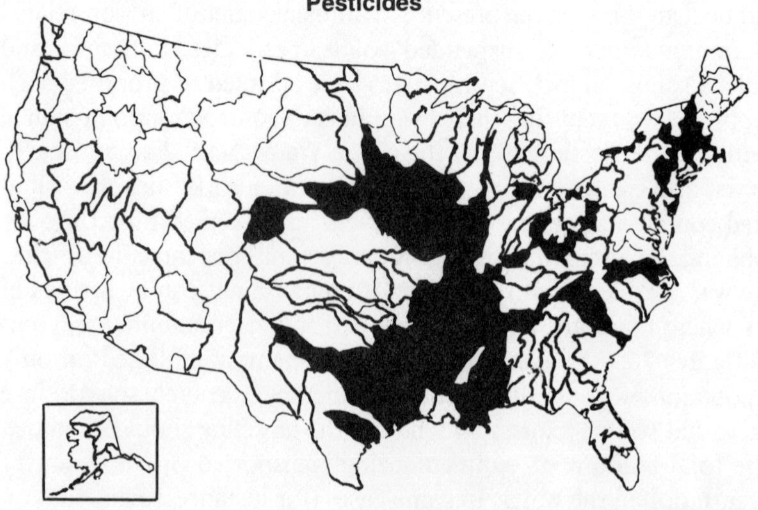

■ Represents basins affected by specific pollutant

Source: U.S. Department of Agriculture, Soil Conservation Service, "Environmental Impact Statement, Rural Clean Water Program" (Washington, D.C.: U.S. Department of Agriculture, 1978), p. 22.

Figure 7.4
Extent of Nonpoint Pollutants—
Nutrients

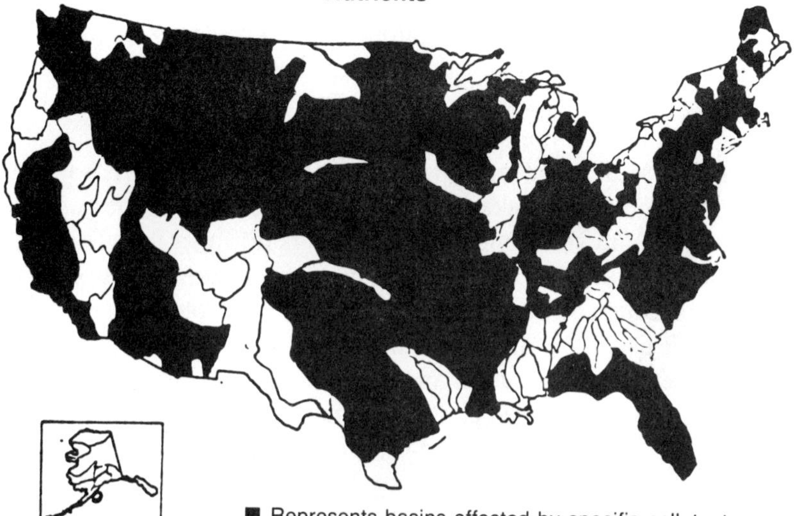

■ Represents basins affected by specific pollutant

Source: U.S. Department of Agriculture, Soil Conservation Service, "Environmental Impact Statement, Rural Clean Water Program" (Washington, D.C.: U.S. Department of Agriculture, 1978), p. 23.

Figure 7.5
Extent of Nonpoint Pollutants—
Dissolved Solids

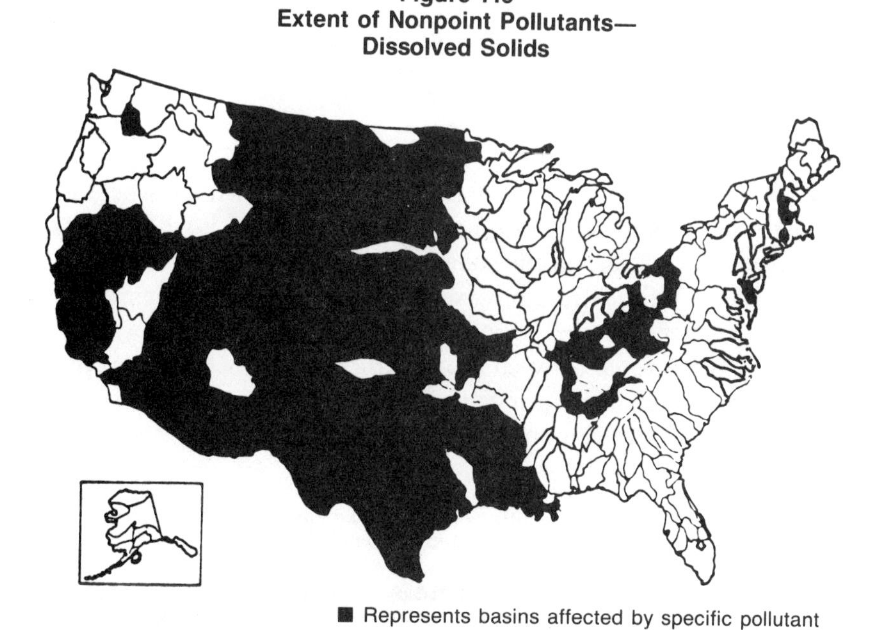

■ Represents basins affected by specific pollutant

Source: U.S. Department of Agriculture, Soil Conservation Service, "Environmental Impact Statement, Rural Clean Water Program" (Washington, D.C.: U.S. Department of Agriculture, 1978), p. 23.

Neither are good water-quality data widely available. The only nationally consistent series is that collected by the U.S. Geological Survey through its National Stream Quality Accounting Network.[41] This system, containing over 500 monitoring stations across the country, might suffice for targeting at the national level, but it does not contain enough monitoring stations to support targeting at the local or, in many cases, even the state level.

Several physical and chemical characteristics of pollution problems related to erosion also make them difficult to monitor properly. Since much erosion and sediment transport occurs during a very few large storms. it would seem to be important to monitor during those storms. However, because most monitoring is done with intermittent "grab samples," it usually is only coincidental if sampling occurs during a high flow period.* In addition, most monitoring efforts rarely include the sediment and adsorbed contaminants that are carried in substantial amounts along a streambed. To correct these problems, water-quality monitoring efforts would need substantial improvement and expansion.†

To be efficient, a targeting scheme must direct erosion-prevention expenditures to areas where they will generate the greatest benefits, but these are not necessarily the areas in which past erosion has created the greatest costs. For instance, as mentioned in chapter 5, if a municipal swimming pool has already been built, reopening an old swimming beach will provide relatively few benefits compared to the cost of building the pool in the first place.

Sources of Pollutants

Even if the federal government becomes able to identify where the off-farm impacts of soil erosion are the most serious, it will still have to gather information on the sources of the pollutants causing those impacts. That is, and probably will remain, the major problem facing efforts to target resources efficiently. Since contaminants may travel a long, tortuous route between the time they leave a field and the time they cause off-site damage, it can be very difficult to determine which particular lands are responsible for an off-farm impact.

---

*In fact, a person responsible for such sampling might intentionally avoid high flow periods for many technical and nontechnical reasons.

†Some of those problems were discovered by states such as Oklahoma when they considered and then rejected such a system for setting priorities on which nonpoint-source pollution problems they should take care of first.[42]

In addition, controlling cropland erosion might not, by itself, eliminate all erosion-related damages. In many areas, a significant portion of the sediment causing problems may come from nonpoint sources such as streambeds that are even more difficult to control than cropland erosion. In such cases, controlling cropland erosion could result in substantial expenditures being made with relatively little downstream improvement.*

Because of these difficulties, most efforts to develop nonpoint-source pollution-control targeting schemes in section-208 plans focused on potential sources of pollutants—that is, the amount of erosion that was occurring—rather than on the impacts that the erosion was causing. Illinois concluded that erosion rates (as estimated by the universal soil-loss equation) provide the only feasible means for deciding which nonpoint-source pollution projects should be undertaken first.[43]

Such an approach is relatively simple. But this advantage is at least partially offset by some serious problems, such as the frequent tendency for methods such as the universal soil-loss equation to be off by a large margin in predicting erosion potential.[44] A second problem is the fact that, as indicated in chapter 2, much of the erosion that does occur may never reach a waterway. In addition, the portion that does reach a waterway can differ substantially from one area to another. Thus, an allocation based on total erosion or on the rate of erosion may not correlate well with the distribution of downstream impacts. Even in its adoption of the universal soil-loss equation as the measure of which erosion-control projects are most necessary, Illinois recognized this fact.

Finally, the amount of, or potential for, erosion may not accurately reflect the relative importance of nutrients and other erosion-related contaminants. Some of these may be more strongly influenced by the quantity of stormwater runoff than the amount of erosion, and most will be strongly related to how much and when fertilizer, manure, and pesticides were applied to a field.

---

*For instance, although erosion and runoff control must be part of any attempt to mitigate eutrophication's harmful effects, eutrophication would continue to be a problem in many waterways even if nutrient runoff could be quickly controlled. This is because the eutrophication process is determined not only by current nutrient inflows but also by a water body's retention of those nutrients. Reservoirs and lakes collect phosphorus and other nutrients from influent streams and store them in sediments that become rich nutrient storehouses. The nutrients can be gradually recycled into the water, perpetuating eutrophication problems for several years after nutrient loads entering the water body have been reduced.

Some states in their section-208 plans proposed other approaches for identifying the most serious pollutant sources. For instance, Michigan, assuming that the seriousness of a pollution problem is related to the amount of erosion entering a stream, adopted a targeting scheme based on the amount of erosion per stream mile in a watershed and used this index to rank all watersheds in the state.[45] It also attempted to adjust for the relative importance of different types of impacts by assigning trout streams a higher priority for remedial action than other streams having the same index value.

North Carolina adopted a more complicated weighting scheme based on the different land uses in each watershed in the state and on the several ways that each type of land use could affect the seriousness of potential pollution problems.[46] Each watershed was assigned a ranking value by combining all these factors. That value theoretically reflected both the seriousness and the extent of potential problems.

Similarly, Oklahoma attempted to develop a relatively sophisticated weighting scheme. However, it was stymied by a lack of data and ended up with a system based on erosion potential that state officials admitted was relatively subjective.[47]

Unfortunately, none of these approaches adequately considers how to accurately identify the sources of erosion-related pollutants. The only efforts that are beginning to deal with this problem are those involved in the development of computer models. A major advantage of models is that they can incorporate all the complicated processes that occur between a farm field and a place where erosion's off-farm impacts occur. Thus, models can consider differing sediment-delivery ratios, sedimentation rates, chemical and biological processes, and other factors that usually are ignored. Ultimately, such models probably represent the only efficient way to target for off-farm effects.

Over the last decade, there has been a rapid development of new computerized mathematical models for analyzing erosion, erosion costs, and nonpoint-source pollution.[48] These models eventually could be very useful in making targeting decisions. Conceptually, at least, a model (or models) could be developed to use readily available information to identify which particular sites are the most important sources of the sediment and other contaminants causing off-farm effects. Such a model also could compare the effectiveness and costs of alternative control measures. And, theoretically, it would be easy and quick to operate, even on microcomputers located in county SCS offices.

Such a model does not now exist. Most current models have been developed for other purposes and cannot be applied to targeting concerns. As one reviewer has observed, "Models are developed for a specific purpose to accomplish a specific job; therefore, application to the model outside specific conditions can result in erroneous answers."[49] Many of the present models are intended for studying individual farms or even specific fields, and most are concerned primarily with identifying the costs and effectiveness of alternative soil conservation measures. Others, though broader in scope, have just not been built to answer targeting questions.

In addition, many of the existing models are very complicated, data intensive, and expensive, although their cost might be more than offset by the money they could save by avoiding misspent resources (assuming that the models could be revised to answer targeting questions).

A major problem is that models are much easier to build than to verify. One theme of a recent conference was that "the 1970s were the age of model development and the 1980s should be the age of model testing. . . . The model development heyday is over, and now we must help facilitate the proper use of these tools and understanding of their capabilities."[50] Although this conclusion may represent a wish more than a prediction, there is no doubt that substantial testing, verification, and modification of the existing models is necessary before they can be counted on to provide the right answers for targeting.

Still, current models can be useful for limited purposes. One set of investigators has used models to make rough estimates of erosion rates, sediment-delivery ratios, and chemical-degradation rates for every watershed in the United States.[51] This information could be used in a national targeting strategy. However, not enough information was available for those investigators to calculate specific rates and ratios for specific watersheds. Instead, their sediment-delivery ratios depended only on regional soil characteristics, with high ratios for clay soils and lower ratios for sandy soils. As a result, their estimates are only rough approximations.

In what seems to be the most ambitious effort yet to model nonpoint-source pollution loadings over a large area, the Ocean Assessment Division of the National Oceanographic and Atmospheric Administration (NOAA) is using a relatively sophisticated model to predict sediment and other nonpoint discharges into U.S. Atlantic coastal waters.[52]*

---

*The ability of models to incorporate many different factors, which is one of their major advantages, can also be a weakness, since often too little is known about these factors to allow anything more than the making of assumptions. These assumptions can be substantially in error but tend to be hidden from individuals using the model. The result can be substantial mistakes made with great precision.

Even when computer models do become able to effectively link the most serious off-site impacts with the lands that provide the responsible pollutants, those models will not necessarily lead to the targeting scheme that results in the greatest reduction of off-farm impacts per dollar spent on soil conservation. To satisfy this efficiency criterion, efforts will have to be focused on those lands where soil conservation techniques will reduce the problem pollutants most economically.

Despite all these problems, some targeting scheme must be adopted soon, while attempts to develop the optimal targeting scheme continue. Without better targeting, the nation's progress in solving the problems of soil erosion's off-farm impacts will continue to be very slow and expensive. But delaying action until the perfect targeting scheme is developed would be even more inefficient. The immediate need is to figure out how to allocate the available resources most efficiently on the basis of current knowledge. At the national level, such a targeting scheme should be based primarily on erosion rates, perhaps modified to take account of differences in sediment-delivery ratios and other factors that could be incorporated with reasonable ease and accuracy.

At the local level, it probably already is possible to target erosion-control expenditures efficiently. The federal government would have to give localities some guidance on how to evaluate the relative importance of different problems, and there would also have to be some mechanisms for insuring that downstream problems were given appropriate consideration and that expenditures were actually made in the proper locations and for the proper purposes. But, if these problems could be overcome, local areas probably have sufficient information to be able to target their resources to those lands where they would most reduce serious problems both locally and downstream.

OTHER IMPLEMENTATION ISSUES

The best strategy for solving a problem often fails because the process for implementing it has not been well thought through. Several questions should be answered before a new program starts: How will the responsibility and authority for carrying out different parts of the strategy be allocated among different agencies and different levels of government? How will the program be monitored? Who will carry out that monitoring? How much money will be needed, and where will it come from? How will the program monitor whether individual landowners are adopting and maintaining the desired techniques? What enforcement actions will be taken if they are not? How much oversight will the different levels of government and the different agencies have over one another? What reporting requirements will there be?

Implementation problems are difficult in any program, but they are likely to be particularly formidable with any strategy to control the off-farm impacts of soil erosion. Millions of separate enterprises would have to be involved in such a program, and, so far, it is not clear which government agency should be given primary responsibility for the strategy. Although agricultural agencies are more able than other agencies to work with farmers, most of them have relatively limited interest and competence in dealing with off-farm concerns. In addition, in a nonpoint-source pollution-control program, they would face a conflict between that program and the basic purpose of most of their other programs—to help farmers increase production.

Water-pollution-control agencies would not face such a potential conflict of interest, but they would be ill-equipped to deal with millions of separate farmers. These agencies already face substantial problems in getting the much smaller number of point sources to comply with existing regulations. Diverting their staffs and attention to address nonpoint problems could make control of those point sources less effective and thereby end up hurting the nation's overall water-pollution-control efforts. In addition, during periods of budgetary constraints, these agencies probably would be unable to acquire sufficient additional resources to implement effective nonpoint-source control problems on their own. Moreover, many of the impacts—for instance, sedimentation in water-supply reservoirs—fall outside these agencies' traditional scope of interest.

It may well be that the responsibility for implementing a targeting strategy needs to be shared by several agencies. A reasonable balance would be for the pollution-control agencies to have primary responsibility for defining needs and strategies in consultation with the agricultural agencies and for the agricultural agencies to have primary responsibility for getting the program implemented with the pollution-control agencies looking over their shoulders. Unfortunately, however, such interagency programs are likely to operate much less efficiently and effectively than those run by a single agency.

The various responsibilities for implementing a control program also must be allocated among different government levels--federal, state, and local. Soil conservation programs have traditionally (and appropriately) been implemented primarily at the local level. Even a large portion of federal employees operate primarily at the local level, providing technical advice to individual farmers. This traditional emphasis on local action has stimulated a large, if uneven, capability at the local level to implement such a program, and some local agencies have already developed imaginative and energetic programs to control nonpoint-source pollution.[53]

In particular, the county-level Soil and Water Conservation Districts have already established many of the communication networks and have much of the information necessary to implement a program to control nonpoint-source pollution. They also have a wide range of authorities that, although often unexercised, could be very useful in implementing an effective erosion-control program.[54]

State governments also will have major roles to play in any agricultural pollution-control strategy. Over the years, many states have instituted programs relevant to controlling off-farm impacts,[55] with many of these programs focusing primarily on assisting local agencies and individual farmers. With a program to control off-farm impacts, state agencies would need to assume somewhat different, additional roles—assisting in decisions on how to allocate funds so they have the greatest impact and stepping in to induce (through enforcement actions, if necessary) recalcitrant landowners to adopt necessary controls. Practically speaking, local agencies could not do the former and often would be unwilling to do the latter.

The federal government will have roles similar to those of the states. In addition, it will need to sponsor whatever research is necessary to support effective programs and to provide technical assistance directly at the local level.

Attempting to implement an effective program through such a complex web of government agencies might make anyone familiar with government operation shudder. Clear priorities must be set. Most immediately, at the federal level there must be a clear statutory statement making nonpoint-source pollution control a primary purpose of existing soil conservation programs. Without such a statement, those programs probably will continue to be managed so as to balance their many competing goals, often accomplishing relatively little with respect to any of them.

CONCLUSIONS

Clearly, developing an effective, efficient program to control off-farm impacts will be difficult. It may not be possible to answer all the necessary questions—certainly not now, if ever. But that does not mean that nothing can be done. With a few significant—but possible—changes in existing erosion- and pollution-control programs, it should be possible to initiate in the near future a program that effectively starts to control erosion's off-farm impacts. Sufficient information, resources, and authority are available. The necessary changes probably could be made with little more than legislative guidance combined with strong, cooperative leadership in the U.S. Environmental Protection Agency and the U.S. Department of Agriculture.

Eventually, for a nonpoint-source pollution-control campaign to be truly effective, the existing resources will have to be supplemented. Regulatory powers will have to be authorized and used. A more accurate targeting scheme—on both a national and local level—will have to be developed. Increased research is especially needed to learn more about where the most serious off-farm impacts are occurring and how they are related to agricultural lands. Still, enough information exists to target efforts toward at least some of the most serious problems. Programs to permanently take the most seriously eroding lands out of row-crop production, or out of production altogether, will have to be initiated, and new farm-management techniques that reduce the amounts of nutrients and pesticides used in agriculture will have to be implemented. Making these changes will be difficult but not impossible. And, even before they are made, much can be accomplished.

In short, the question is not whether an effective program to control the off-farm impacts of soil erosion is possible. Rather, the question is whether U.S. society has the will.

REFERENCES

1. An estimate of $18 billion is provided in U.S. Comptroller General, *Agriculture's Soil Conservation Programs Miss Full Potential in the Fight against Soil Erosion*, Report to the Congress, GAO/RCED-84-48 (Washington, D.C.: U.S. General Accounting Office, November 28, 1983), p. i; but Tom Fulton and Peter Braestrup, "The New Issues: Land, Water, Energy," *The Wilson Quarterly* 5, no. 3 (1981):122, estimate up to $30 billion as of 1977.

2. P. J. Nowak and P. F. Korsching, "Social and Institutional Factors Affecting the Adoption and Maintenance of Agricultural BMPs," in Frank W. Schaller and George W. Bailey, eds., *Agricultural Management and Water Quality* (Ames, Iowa: Iowa State University Press, 1983); Ted L. Napier et al., "Factors Affecting the Adoption of Conventional and Soil Conservation Tillage Practices in Ohio" (accepted for publication by *Journal of Soil and Water Conservation*); Waldon R. Kerns et al., "Nonpoint Source Management: A Case Study of Farmers' Opinions and Policy Analysis" (Blacksburg, Va.: Virginia Polytechnic Institute and State University, n.d.); Lee A. Christensen and Patricia E. Norris, "Soil Conservation and Water Quality Improvement: What Farmers Think," *Journal of Soil and Water Conservation* 38, no. 1 (1983); Lee A. Christensen and John A. Miranowski, "Perceptions, Attitudes and Risks: Overlooked Variables in Formulating Public Policy on Soil Conservation and Water Quality—An Organized Symposium," prepared for U.S. Department of Agriculture, Economic Research Service, ERS Staff Report no. AGES820129, 1982; *Water Resources Bulletin* 19, no. 3 (1983):459.

3. Ohio Environmental Protection Agency, Ohio Water Quality Management Plan, Office of the Planning Coordinator, "Scioto River Basin Agriculture Report," May 1981, revised January 1983; and Frank Clearfield, "Nonpoint Source Pollution—A People Problem," *Extension Review* 54 (Spring 1983):46.

4. U.S. Comptroller General, *Agriculture's Soil Conservation Programs Miss Full Potential in the Fight Against Soil Erosion*, pp. 22, 34.

5. Clayton W. Ogg, James D. Johnson, and Kenneth C. Clayton, "A Policy Option for Targeting Soil Conservation Expenditures," *Journal of Soil and Water Conservation* 37, no. 2 (1982); U.S. Department of Agriculture, Agricultural Stabilization and Conservation Service, *National Summary Evaluation of the Agricultural Conservation Program, Phase I* (Washington, D.C.: U.S. Department of Agriculture, 1978), p. 18 and calculated from information in table 8, p. 26; and U.S. Comptroller General, *To Protect Tomorrow's Food Supply, Soil Conservation Needs Priority Attention* (Washington, D.C.: U.S. General Accounting Office, 1977), p. 3.

6. U.S. Comptroller General, *Agriculture's Soil Conservation Programs Miss Full Potential in the Fight Against Soil Erosion*, pp. 41-42.

7. J. M. McGrann and J. Meyer, "Farm-Level Economic Evaluation of Erosion Control and Reduced Chemical Use in Iowa," in Raymond C. Loehr et al., eds., *Best Management Practices for Agriculture and Silviculture* (Proceedings of the 1978 Cornell Agricultural Waste Management Conference) (Ann Arbor, Mich.: Ann Arbor Science Publishers, 1979), p. 368; and W. D. Seitz, C. Osteen, and M. C. Nelson, "Economic Impacts of Policies to Control Erosion and Sedimentation in Illinois and Other Corn-Belt States," in Loehr et al., *Best Management Practices for Agriculture and Silviculture*, p. 376.

8. James J. Jacobs and George L. Casler, "Internalizing Externalities of Phosphorus Discharges from Crop Production to Surface Water: Effluent Taxes vs. Uniform Reductions," *American Journal of Agricultural Economics* 61 (1979):309-12.

9. Ken Cook, "Cross-Compliance: Is It Bold, Menacing or Just Plain Dumb?" *Journal of Soil and Water Conservation* 39, no. 4 (1984):250-51.

10. Some of the programs that might be included at the federal level in a cross-compliance program are discussed in U.S. Department of Agriculture, *RCA Program Report and Environmental Impact Statement*, review draft (Washington, D.C.: U.S. Government Printing Office, 1980). Other programs are discussed in Dean T. Massey and Margaret B. Silver, "Property Tax Incentives for Implementing Soil Conservation Programs under Constitutional Taxing Limitations," *Denver Law Journal* 59, no. 3 (1981).

11. Clayton W. Ogg, Arnold B. Miller, and Kenneth C. Clayton, "Agricultural Program Integration to Achieve Soil Conservation," n.d., p. 10.

12. John F. Timmons and Dennis C. Cory, "Responsiveness of Soil Erosion Losses in the Corn Belt to Increased Demands for Agricultural Products," *Journal of Soil and Water Conservation* 33, no. 5 (1978):221-26.

13. Ogg, Johnson, and Clayton, "A Policy Option for Targeting Soil Conservation Expenditures," p. 69.

14. Strong reactions are also expressed in Congress. See Christopher Leman, "Political Dilemmas in Evaluating and Budgeting Soil Conservation Programs: The RCA Process," in Harold G. Halcrow, Earl O. Heady, and Melvin L. Cotner, eds., *Soil Conservation Policies, Institutions, and Incentives*, prepared for North

Central Research Committee III (Ankeny, Iowa: Soil Conservation Society of America, 1982), p. 66.

15. Christensen and Norris, "Soil Conservation and Water Quality Improvement," p. 17.

16. See Massey and Silver, "Property Tax Incentives for Implementing Soil Conservation Programs under Constitutional Taxing Limitations"; Wayne D. Rasmussen, "History of Soil Conservation, Institutions and Incentives," in Halcrow, Heady, and Cotner, *Soil Conservation Policies, Institutions, and Incentives*, pp. 16-17; and Winston Harrington et al., "Assessment of Nonpoint Source Pollution Control Policies," Resources for the Future, Washington, D.C., March 1984, p. 12, for a discussion of some of these.

17. Sandra S. Batie, *Soil Erosion: Crisis in America's Croplands?* (Washington, D.C.: The Conservation Foundation, 1983), pp. 122, 123.

18. Wayne D. Rasmussen, "History of Soil Conservation, Institutions and Incentives," in Halcrow, Heady, and Cotner, *Soil Conservation Policies, Institutions, and Incentives*, p. 16.

19. Daniel W. Bromley, "The Rights of Society vs. the Rights of Landowners and Operators," in Halcrow, Heady, and Cotner, *Soil Conservation Policies, Institutions, and Incentives*, p. 220.

20. James L. Arts, "Private Property and Soil Loss Regulations," *Journal of Soil and Water Conservation* 36, no. 6 (1981):317-19; and Beatrice H. Holmes, "Institutional Bases for Control of Nonpoint Source Pollution under the Clean Water Act—with Emphasis on Agricultural Nonpoint Sources" (Washington, D.C.: U.S. Department of Agriculture, 1979).

21. Lawrence W. Libby, "Interactions of RCA with State and Local Conservation Programs," in Halcrow, Heady, and Cotner, *Soil Conservation Policies, Institutions, and Incentives*, p. 122; and Holmes, "Institutional Bases for Control of Nonpoint Source Pollution under the Clean Water Act."

22. Sandra S. Batie, "Policies, Institutions and Incentives for Soil Conservation," in Halcrow, Heady, and Cotner, *Soil Conservation Policies, Institutions, and Incentives,* p. 32.

23. Timothy R. Henderson, "Mandatory Regulation of Soil Erosion on Agricultural Lands: State Laws and their Implementation" (Master's thesis, University of Wisconsin, 1980); and Holmes, "Institutional Bases for Control of Nonpoint Source Pollution under the Clean Water Act."

24. Ted L. Napier and D. Lynn Forster, "Farmer Attitudes and Behavior Associated with Soil Erosion Control," in Halcrow, Heady, and Cotner, *Soil Conservation Policies, Institutions, and Incentives*, p. 148.

25. American Farmland Trust, *Soil Conservation in America: What Do We Have to Lose?* (Washington, D.C.: American Farmland Trust, 1984).

26. New Hampshire Water Supply and Pollution Control Commission, *Interim Report—Sediment and Erosion Control* (Concord, N.H.: New Hampshire Water Supply and Pollution Control Commission, 1979), p. 7.3-1.

27. National Association of Conservation Districts, *Tuesday Letter,* December 13, 1983, p. 2.

28. Harold G. Halcrow and Wesley D. Seitz, "Soil Conservation Policies, Institutions and Incentives: A Summary," in Max Schnepf, ed., *Natural Resources Policy: Research Strategies for the Future* (proceedings of a conference) (Ankeny, Iowa: Soil Conservation Society of America, n.d.), p. 9; and American Farmland

Trust, *Soil Conservation in America.*

29. Ohio Environmental Protection Agency, "Scioto River Basin Agriculture Report," p. 107.

30. Napier and Forster, "Farmer Attitudes and Behavior Associated with Soil Erosion Control," p. 148.; "Impacts of Sediment on Biota in Surface Waters," in U.S. Environmental Protection Agency, Environmental Research Laboratory, *Impacts of Sediment and Nutrients on Biota in Surface Waters of the United States*, EPA-600/3-79-105 (Athens, Ga.: Environmental Protection Agency, 1979); Robert D. Ohmart, Wayne O. Deacon, and Sten J. Freeland, "Dynamics of Marsh Land Formation and Succession Along the Lower Colorado River and Their Importance and Management Problems as Related to Wildlife in the Arid Southwest," in *Transactions—North American Wildlife and Water Resources Conference* (40th annual conference, 1975), pp. 240-54; and J. M. Rice, "Management and Financing of Agricultural BMPs," in Loehr et al., *Best Management Practices for Agriculture and Silviculture*, p. 335.

31. The Conservation Foundation, *State of the Environment 1982* (Washington, D.C.: The Conservation Foundation, 1982), p. 293.

32. U.S. Department of Agriculture, *1980 Appraisal Part I: Soil, Water, and Related Resources in the United States—Status, Conditions, and Trends* (Washington, D.C.: U.S. Government Printing Office, 1981), pp. 73, 156.

33. U. S. Comptroller General, *Agriculture's Soil Conservation Programs Miss Full Potential in the Fight against Soil Erosion*, pp. 22, 34.

34. U.S. Department of Agriculture, Agricultural Stabilization and Conservation Service, *National Summary Evaluation of the Agricultural Conservation Program, Phase I.*

35. Ogg, Miller, and Clayton, "Agricultural Program Integration to Achieve Soil Conservation," pp. 14, 15, quoting a study by the U.S. Department of Agriculture.

36. Norman A. Berg and Robert J. Gray, "Soil Conservation: 'The Search for Solutions,' *Journal of Soil and Water Conservation* 39, no. 1 (1984).

37. U.S. Comptroller General, *Agriculture's Soil Conservation Programs Miss Full Potential in the Fight Against Soil Erosion.*

38. U.S. Department of Agriculture, Soil Conservation Service, Conservation Planning and Application Staff, State and Local Operations, "Conservation Technical Assistance National Targeting Plan, Fiscal Years 1981–84" September 1982, p. 5.

39. U.S. Comptroller General, *Agriculture's Soil Conservation Programs Miss Full Potential in the Fight Against Soil Erosion*, p. 29.

40. Oregon Department of Environmental Quality, Water Quality Program, "Oregon's Statewide Assessment of Nonpoint Source Problems: Plate 2—Sedimentation, August 1978.

41. Richard A. Smith and Richard B. Alexander, "A Statistical Summary of Data from the U.S. Geological Survey's National Water Quality Networks," prepared for the U.S. Department of the Interior, Geological Survey, Open File Report no. 83-533, June 1983.

42. Greer County Conservation District, "Final Report: Update Nonpoint Source Ranking," prepared for the Oklahoma Conservation Commission, Task no. 134, FY 78/79 208 Work Plan; and Oklahoma Contributing Areas, "Program Design for Water Quality Monitoring and Watershed Data Update," Letter Report, Task no. 705.

43. Northeastern Illinois Planning Commission, "Establishing Priorities for Implementing Agricultural Best Management Practices," Areawide Clean Water Planning Staff Paper, prepared for U.S. Environmental Protection Agency, March 1980, p. 7-1.

44. David B. Baker, "Fluvial Transport and Processing of Sediment and Nutrients in Large Agricultural River Basins," prepared for Lake Erie Wastewater Management Study, February 1982, p. 114.

45. Michigan Department of Natural Resources, Water Quality Division; Michigan Department of Agriculture, Soil and Water Conservation Division; and Michigan State University, Department of Resource Development, "An Agricultural Strategy for Reducing Nonpoint Source Pollution in Michigan," March 1982, pp. 14-19.

46. North Carolina Department of Natural Resources and Community Development, Division of Enviromental Management, "208 Phase 1 Results: Ranking of 128 Sub-basins for Water Pollution Potential based on Intensity of Land-Disturbing Activities," January 1978, p. C-1.

47. Greer County Conservation District, "Final Report: Update Nonpoint Source Ranking."

48. D. R. Keeney, "Transformations and Transport of Nitrogen," and other papers in the same volume, Frank W. Schaller and George W. Bailey, eds., *Agricultural Management and Water Quality* (Ames, Iowa: Iowa State University Press, 1983), pp. 48-64.

49. Michael K. Reichert, "Use of Selected Macroinvertebrates as Indicators of Sedimentation Effects on Huntington River, Utah," (Master's thesis, Brigham Young University, 1975), p. 179.

50. F. J. Humenik et al., "Conference Summary and Research Needs: An Environmental Perspective," in Schaller and Bailey, *Agricultural Management and Water Quality*, p. 459

51. L. P. Gianessi, "National Modeling and Policy Analysis of Agricultural Nonpoint Source Control Options," in Schaller and Bailey, *Agricultural Management and Water Quality*, pp. 403-25.

52. Personal communication from Dan Busta, Ocean Assessment Division, National Oceanic and Atmospheric Administration.

53. A. Weinberg and Jim Arts, *Local Government Options for Controlling Nonpoint Source Pollution* (Madison, Wis.: University of Wisconsin-Extension, 1981); and U.S. Environmental Protection Agency, Office of Great Lakes National Programs, "Voluntary and Regulatory Approaches for Nonpoint Source Pollution Control (Water Quality Planning)," Proceedings of a conference held at Rosemont, Ill., May 23-24, 1978, pp. 42-47.

54. Dean T. Massey, "Land Use Regulatory Power of Conservation Districts in the Midwestern States for Controlling Nonpoint Source Pollutants," *Drake Law Review* 33, no. 1 (1983–1984); and *Farmland Notes*, July 1984.

55. Allen L. Fisk and Carole Richmond, "Soil Conservation and Farmland Protection: The Western Perspective," discussion draft, prepared in cooperation by the Western Governors' Association, the Western Office of the Council of State Governments, and the Lincoln Institute of Land Policy, May 1984, details the various relevant programs in western states. Similar programs exist in most other states as well and are often described in the state section-208 water-quality plans.

Index

Italicized numbers represent figures, and "n" numbers are for footnotes.